The WANDERING GENE

and the INDIAN PRINCESS

ALSO BY JEFF WHEELWRIGHT

Degrees of Disaster (1994)

The Irritable Heart (2001)

The WANDERING GENE
and the INDIAN PRINCESS

RACE, RELIGION, AND DNA

Jeff Wheelwright

W. W. NORTON & COMPANY

New York London

To protect the privacy of two men who married into the Medina family
and who lost their wives, I have changed their names.
Bill Kramer and Michael are pseudonyms.

For information about permissions to reproduce selections from this book, write to
Permissions, W. W. Norton & Company, Inc., 500 Fifth Avenue, New York, NY 10110

For information about special discounts for bulk purchases, please contact
W. W. Norton Special Sales at specialsales@wwnorton.com or 800-233-4830

Manufacturing by RR Donnelley, Harrisonburg
Book design by Dana Sloan
Production manager: Devon Zahn

Library of Congress Cataloging-in-Publication Data

Wheelwright, Jeff.
The wandering gene and the Indian princess : race, religion, and DNA / Jeff Wheelwright.
p. ; cm.
Includes bibliographical references and index.
ISBN 978-0-393-08191-6 (hardcover)
I. Title.
[DNLM: 1. Breast Neoplasms—genetics. 2. Genetic Predisposition
to Disease. 3. Ethnic Groups—genetics. 4. Genes, BRCA1.
5. History of Medicine. 6. Religion. WP 870]
LC classification not assigned
616.99'449042—dc23
2011030178

ISBN 978-0-393-08191-6

W. W. Norton & Company, Inc., 500 Fifth Avenue, New York, N.Y. 10110
www.wwnorton.com

W. W. Norton & Company Ltd., Castle House, 75/76 Wells Street, London W1T 3QT

1 2 3 4 5 6 7 8 9 0

In memory of Dale Owen

Everything is foreseen, yet free will is granted; the world is ruled with divine goodness, yet all is according to the greatness of one's deeds.

—PIRKEI AVOT (*ETHICS OF THE FATHERS*)

CONTENTS

PROLOGUE

S honnie Medina was a happy girl. She was a happy girl who felt she would die young. But all that was under the surface. On the surface she was beautiful.

Her physical beauty, when she was a young woman in Culebra and a young wife in Alamosa, was the primary thing that people mentioned about her. Photographs and snatches of videotape don't quite capture it because fundamentally what people were talking about was charisma. It came through her looks when she was in front of you, tossing her full head of dark hair and giving you her full attention. Then her beauty acted like a mooring for her other outward qualities, undulating from that holdfast like fronds of kelp on the sea. Then Shonnie was magnetic, vain, kind to others, religious without reservation, funny, a little goofy, and headstrong.

Being headstrong or unreasonable was the quality that the doctors in Alamosa and Denver blamed for her death—for Shonnie was right about dying young. She carried in her cells a dangerous genetic mutation and died when she was just twenty-eight, having refused surgery for her aggressive, inherited breast cancer.

The gene she carried is known as BRCA1.185delAG, a famous gene with a long pedigree. This book will tell the story of the gene, with background about the scientists and other professionals who have grappled with that crucial stretch of DNA. The book will also tell of Shonnie's large family in the southern part of Colorado's San Luis Valley, radiating from their genetic heritage to questions of faith and identity. The book will boil down to a medical and spiritual choice. Shonnie Medina might

1

have saved herself if she'd put herself into the hands of scientific medicine. Jealous of her body, oblivious to the gene, she insisted on another style of care.

The happiest day in her wholesome life was her wedding day: August 1, 1992. Shonnie was twenty-two. To marry Michael, an Anglo boy from Alamosa, Shonnie made something of a cultural leap. The Medinas are Hispanos, a mix of Spanish and Indian people. Older than other Hispanics in North America, the Hispanos claim a four-hundred-year history in northern New Mexico and southern Colorado. Their villages, dotting the northern reach of the Rio Grande, were as lively and insular as shtetls. Culebra itself was a tight cluster of villages where everybody knew everybody else, and for that reason Marianne Medina, Shonnie's mother, was concerned about the all-too-close genetic relationships. She told her daughter, You're going to marry someone from away from here, thinking more about geography than about race. Shonnie did, although it was only fifty miles north and not the Hispano type that her mother expected.

If there was a racial element to her identity, it's hard to describe. For one thing, Hispanos regarded themselves as an ethnic group, not a race. For another, Shonnie's religion had taught her to be color-blind. She hardly spoke of race, other than to make light of Indianlike misspellings of her name—when, for instance, she received letters addressed to Shawnee. Her complexion was very fair. For centuries the Hispanos of New Mexico had prized the pure pale skin of the *españole* and had strived to upgrade their racial status and skin tone through marriage. Yet Shonnie Medina tanned deeply and was notorious for showing off her tan, spurning the weak winter light of Culebra. She dared to look dark, Native American dark, because of her confidence in her whiteness. Indeed, a few days before the wedding she decided that Michael looked too pale next to her. She had him undergo a stiff treatment with tanning lotion—which turned his face orange and prompted a frantic scrubbing. Whatever Shonnie thought about her racial or ethnic heritage in America, her wedding would make her assimilation complete. For all that, a geneticist would be able to look below her skin and easily distinguish the Hispano from the Anglo she was marrying.

On the day of the wedding, it was warm and not windy in Culebra. The high prairie sky was clear. Chairs were set up in the gravel parking lot facing the Kingdom Hall, and a low stage was prepared for the young people of the nuptial party, separate from the rest of the gathering. While the groomsmen directed the arriving cars to park around the back, the afternoon's clouds took their seats on the front row of the mountains. There were more than a hundred guests, a few of whom put up umbrellas against the sun.

The procession began, to the accompaniment of the taped melodies that are prescribed for such occasions. Steps crunching on gravel, the somber ushers and colorful bridesmaids walked down the aisle between the chairs. Marianne Medina had made the girls' dresses, all but one, and she didn't stint on the bright fabric. Here comes Veronica Sanchez on the arm of an usher. Veronica is Shonnie's first cousin once removed. Her grandmother and aunt died of the family disease. Here's Jackie Van Geison, whom Shonnie has been instructing in the Bible. Jackie will meet her future husband here today. And so on, until the exotic bride and her father appear, brilliant in the open air.

Shonnie had seen a picture in a magazine of an Indian (Asian Indian) bridal gown. Judging from Marianne's adaptation of it, the gown must have featured mounds of satin and a multitude of jewels. Shonnie's dress glinted with sequins and dripped with beads and teardrop pearls. The skirt was slit down one side. The back had a bustle, a detachable bow, and a long, flouncy train. Although Marianne had lined the underside of the train with a protective strip so that it could drag without harm, for most of the day Shonnie carried the train on the crook of her arm, or else she hung it from a special strap on her wrist, which spoiled the effect, Marianne thought.

Joseph, Shonnie's father, had vetoed the bodice that was pictured in the bridal magazine because he thought its heart-shaped bustline was too revealing. Instead, Shonnie went with a lacy, scalloped pattern across her collarbone. Her wedding veil could barely contain her sumptuous black hair. Her nails were manicured with French tips. The pièce de résistance was a jeweled headband, which she wore low on her forehead, Apache-style, its stones having been taken from her grandmother Dorothy's necklace. Taller in heels than her father, Shonnie passed by her guests smiling, and Joseph, escorting her, wore a huge grin. When they reached the platform and Joseph let her go, the groom picked up the bridal train and the two mounted to their seats. The attendants were already sitting.

The officiating elder, who happened to be Michael's father, began to explain the marital contract to the young people. In an address lasting forty minutes, the elder covered the responsibilities and requirements of husband and wife as stipulated by scripture. Two are better than one, advises Ecclesiastes (4:9). The husband, designated the head, is to be respectful of the wife, the elder said, not tyrannical. Her honor is precious, delicate, cherished. Leaning toward Shonnie, Michael held open a Bible for both of them and expertly turned the pages to the passages being cited. The gesture symbolized that, though the husband directed the couple's spiritual life, the two were literally on the same page. It would be interesting to know what Shonnie was thinking then.

Ecclesiastes goes on to observe that if two lie down together, they will certainly get warm. That Shonnie and Michael were virgins did not need to be stated. Probably their bodies had not warmed one another for more than a few seconds; they had resisted what were called the desires incidental to youth. In fact, they had rarely been alone because the orthodox youth of their circle socialized together, usually under the eye of a chaperone.

At last they spoke their marriage vows. Michael took the microphone first, and then Shonnie. Her voice was succinct and soft. Among the onlookers, only Shonnie's sister did not look happy. Iona was losing her big sister, her only sibling, for the first but not the last time. Instead of ending with "So help me God" or the like, the vow was sealed, rather woodenly, with "According to God's marital arrangement." Husband and wife turned and presented themselves to their witnesses. They kissed and walked shyly off the platform, up the aisle.

The reception took place at the high school gymnasium. Spanish fans decorated the steep banks of the folded bleachers. Removing her veil,

Shonnie extended a hand to her guests. Out of the competing glare of the sun she looked even more luminous. George Casias, the wedding photographer, estimated that four hundred people came. The two families, because they shared a faith, mingled comfortably. Michael's side did find it unusual that the newlyweds opened their presents in the middle of the reception, but that's how Hispanos did it.

The couple's first dance was to a recording of a slow tune, "Love of a Lifetime." Shonnie rested her arms on Michael's shoulders and enfolded the nape of his neck with her hands. The white tips of her nails looked like whole notes swaying on a bar of music. His entire right side was covered by her train, as if by a swan's wing. After Joseph danced with her, she and Michael cut the wedding cake. It was an elaborate, almost garish, four-tiered affair, erected over a miniature fountain and held firm by struts on two sides. Linking arms, Shonnie and Michael drank iced tea from plastic champagne glasses, someone having forgotten the ceremonial wine, though this was a dry party anyway.

"Hey Baby, Que Paso?" played next, their fast dance ("I thought I was your only *vato*"). Whether the beat was fast or slow, Shonnie was an instinctive, supple dancer, he not so much, to put it kindly. When Americans first traveled over the Santa Fe Trail to New Mexico, they remarked on the fondness for dancing among this strange, uninhibited, Catholic people. Light in the arms of the horny *gringos*, the ladies would tip back their heads and make merry laughter, all the while smoking their corn-husk cigarettes.

At 7 p.m. it was time for the newlyweds to leave. On the videotape, dusk transfigures the summer sky, the orange fingers of the departed sun still harrying the diaspora of cloud. Michael carries his wife, her head averted, to the front of the high school. He tells Shonnie to keep her eyes closed until he stops. The surprise is a waiting white limousine. All the Culebrans are astonished, because no one has ever furnished a limo at a Spanish wedding before.

Now they pose for one more picture; now they raise their champagne glasses to their lips with arms interlocked. At the send-off Shonnie doesn't toss her garter to her bridesmaids. That would be too racy a gesture for a woman like her. Fishing her car keys from a compartment of her dress, she flips them—how many times has Shonnie forgotten or lost her keys?—to her sister. The dashing couple turns, waves, and disappears into the limo with the quick agile grace of their youth. The gene gets in with her, the gene that has followed her from Judea to Sepharad to Mexico and up the winding aisle of the Rio Grande.

There is a passage in the adventures of Don Quixote where the addled knight comes briefly to his senses. His sidekick, Sancho, informs him that the lady Dulcinea, the inspiration for Quixote's quest, is not a noblewoman after all but a lowly maid of the village.

So what? Quixote declares. So what if I apostrophize her? He concludes, You should know, Sancho, if you don't know it already, that the two qualities above all others that inspire love are beauty and reputation. And these two Dulcinea is in consummate possession of.

Chapter 1

GIRASOL

En un corral redondo	In a round corral
con vacas en el fondo	With cows far back
y un pastor hermoso	Are a beautiful shepherd
y un perro rabioso	And a rabid dog

—RIDDLE FROM OLD NEW MEXICO

To cross the border from New Mexico into the San Luis Valley of Colorado is to cross a meaningless line. State highway 522 becomes state highway 159, and Taos County Costilla County. The run-down village of Costilla, New Mexico, becomes the run-down village of Garcia, Colorado. By all that is holy and true, the way is still New Mexico, at least as far as Alamosa.

To the west is the clean-edged cone of Ute Mountain, incongruously standing by itself on the plain that widens to the San Luis Valley. On the right, growing more imposing with each mile, is the Sangre de Cristo range, which encloses the east side of the Valley. These mountains are said to have been named by a dying man, Francisco Torres. Part of Torres had already died before he spoke his famous last words. Of noble Spanish ancestry, having been engaged to be married in Seville and suffering the death of his fiancée only days before the wedding, the heartbroken man retired to a monastery, whereupon he learned of an expedition to the northern frontier of New Mexico. (How many powerful thoughts have been hatched in the safety of monasteries!)

So off he went, the reconstituted Fray Torres. Two bony ranges, the Sangre de Cristos and the San Juans, join at the northern perimeter of the Valley and form a sort of rib cage opening onto New Mexico. After reaching the source of the Rio Grande River in the San Juan Mountains, the Spaniards made their *entrada* to the San Luis Valley from the west. All the while, they were being hounded by Indians. Fray Torres was badly wounded in one of the attacks, and the party had to hole up at the foot of the eastern mountains. On this August day of 2007, the prairie is rather hazy and the peaks a vague gray, so that the sunset will have less to work with, but at the time Fray Torres lay bleeding, snow draped the top of *la sierra* like a waiting canvas. As seen from the valley floor, the darkening red in the clefts of the mountains would have flushed to the sky like magma. One story has it that Torres got to his feet in a kind of ecstasy, crying, *Sangre de Cristo! Sangre de Cristo!* [Blood of Christ!], before collapsing. In another version, he lay there watching the effect, and whispered the words.

The elevation, 7,500 feet, is noticeable to any visitor from sea level. Lungs pinched, the northbound traveler turns instinctively toward the shelter and water of the mountains, away from the exposure and aridity of the prairie. Near the state line, Costilla Creek flows busily out of the Sangre de Cristos and intersects the highway. After driving up the creek a few miles, the water sporting over boulders, you come to a colorful cemetery. It has small American flags stuck on the graves, and pinwheels apparently, or possibly Pueblo Indian prayer-sticks?—no, they are gaudily adorned crosses whose arms are fixed. Naked to the sky, the Santo Niño Catholic cemetery is without greensward or landscaping, just mineral soil and dusty weeds separating the graves.

The detour continues east along Costilla Creek, past vacation ranches on the floodplain and trout-fishing spots in the national forest, but those are Anglo elaborations on the landscape and would break the spell. Instead, turn north again at the cemetery, mount the dry tableland (still in the cool aura of the mountains), and cross into the Culebra Creek drainage of the San Luis Valley. You have reached the perimeter of the Spanish colonization from Mexico. *Rio Culebra*, the serpent river. Via sinuous channels and man-made tangents, Culebra Creek and its tributaries irrigate the southeast corner of the Valley, where the Medina family lives and where this story takes place, this story of genes and faith. The people who settled here had Indian blood in their veins—they had Indianness circulating within their Catholic Spanishness, whether they admitted it or not.

Ever since Santa Fe, the sunflowers have been mesmerizing. Bisecting the sagebrush and rabbitbrush, brilliant yellow sunflowers line both sides of the road and project onto the horizon a single yellow vanishing point, which holds your attention like a hood ornament. With their purposeful heads and vaguely vertebrate forms, the sunflowers could constitute a race of people marching north. The common sunflower, *Helianthus annuus*, is a camp follower of human beings. Native Americans were the first to domesticate the plant and distribute it. It traveled to Europe with homecoming Spanish explorers and returned to America a few centuries later, much modified, in the company of German, Russian, and other immigrants. The cultivated strains are put to many uses, both industrial and horticultural. Meanwhile, the feral common sunflower, which never left, sprouts on the roadsides and in fallow fields, and is considered a weed in several states.

Along roads in San Luis Valley you can find scraggly, foot-tall sunflowers, their leaves wizened, barely flowering in cracks of asphalt, while a few miles away, lush six-footers are hulking on a floodplain of Culebra Creek. They are of the same race. More than most plants, the common sunflower has phenotypic plasticity, which means a potential for a plethora of shapes. Although the environment (obvious differences in soil, light, water, nutrients) can account for the difference in form, geneticists have wondered about the sunflower's DNA. After all, it is the genes of the plant that tell the plant what to do in response to the environment. The sunflower, scientists say, is genetically versatile, its DNA so full of variation that it can grow in any reasonable habitat. A population of sunflowers appears to contain templates for a host of identities, as if an awesome package of environmental software were preloaded onto the hard drive of the genes. Global warming? Renewed ice age? Bring it on; the sunflower will adjust and assimilate.

Its plasticity is thought to have developed because during its evolutionary past the plant bred with other types of sunflowers in North America and picked up from them useful varieties of critical genes. In this region the nearest relative to the common sunflower is the prairie sunflower, *H. petiolaris*. It can be distinguished from the other by its narrower, more lance-shaped leaves and its preference for sandy soil. Also, there's a tall but small-blossomed sunflower that you see occasionally in people's gardens. It differs from the rest in having an underground identity, an edible tuber, hence the scientific name *H. tuberosus*. This sunflower too was taken from America to Europe, and when it returned, it was known as the Jerusalem artichoke or Jewish potato. In fact, Marianne Medina, in Culebra, had the impression that Jewish potato was the nickname for all sunflowers in general. Marianne's uncle used to grow Jewish potatoes, uprooting them in the fall and saving some of the knobby tubers to start the next year's crop. A proud and hardy perennial, Jewish potato will not mingle its genes with the genes of an annual sunflower, even when the two are grown side by side.

At any rate, it is easy to get carried away with genes when so much progress is being reported today in genetics. The idea of DNA has become so familiar that the metaphor may mean more than the material. To say that something is in your DNA is to enunciate a defining element of your character, an unshakable part of your core. But lurking below the metaphor is the bigger and darker idea known as genetic determinism, the belief that a person's life is prefigured in her genetic inheritance without her being able to do anything about it. Here is an unromantic picture of genes riding roughshod over the landscapes of human potential. Genetic determinism implies that, at best, your DNA hands you a circumscribed or pigeonholed existence, and, at worst, a tragically truncated life. The original and still most powerful basis for genetic determinism is heritable disease, the kind that cut into the Medina clan.

Still, the outcomes of a life are not predestined in the biochemical lettering of DNA. The genetic text of a plant or a person is not determinative except in those very rare instances where a vital gene is marred beyond repair. Rather, the *landscape* evokes the life, the particulars of the landscape. The DNA of a seed, reacting to the world where the seed fell, is directed down one path of development or another. What better illustration of the commanding role of the environment than the sun and sunflower, the plant faithfully twisting its head to follow the arc of the star? Likewise, a human being starts to be molded by the outside world even before exiting the womb and experiencing the light. In this depiction of the nature–nurture question, the landscape informs the life, and simultaneous other lives are implicit in the roads not taken, in the environments where life did not happen to take root.

So the environment of northern New Mexico—the thin air, the mesas and sagebrush, the mountains filing by, the cold, captured creeks—surely evoked special traits in the people who settled here. Reciprocally, a plaintive Spanishness and a primitive Catholicism seeped into the sandstone hills and mingled with the mute substrate laid down by the Indians. Almost everywhere else in the region, the ore of Old New Mexico has been buried by progress, but its seams still lie on the surface in the San Luis Valley, and particularly in Culebra. Here, in this sprinkle of Hispano villages hard by *la sierra*, Shonnie Medina inherited a genetic mutation older than Christ (and from the same place as Christ) that determined her life. In this meeting of nature and nurture, the double helix ignored the coaxing of the landscape.

When summer clouds appear in the morning over the Valley, they tend toward streaks and smears. As the day decides what to do, southwest winds make the clouds clump and swell with moisture. On this August afternoon, more humid than the day Shonnie married, cumulus clouds are gathering to the east. The Sangre de Cristo summons fluffy clouds to its crest like a shepherd calling for his sheep, and already there is some grumbling on high. Either it will rain hard where you are standing or not at all, for the cloudbursts are brief and focal in an alpine desert that receives only eight inches of precipitation a year. Following the rain or not, the clouds flatten out again and float more freely, backing away from the mountains and mottling the sky. For the original Puebloans, clouds and lightning were symbols of male fertility (women's fertility being earth- and crop-centered) because of the way that clouds built to a climax and groaningly released. Clouds replenished the serpentine creeks. On quite another level, clouds represented the spirits of the dead. It was a comfort to everyone when clouds appeared.

As you come down off the tableland, Culebra presents itself as a shadow-spattered mosaic of green rectangles. The rectangles are fields whose edges contain water; the water is being shunted from the creeks by a colonial-era system of canals, or *acequias*. On your right is San Francisco Creek, a tributary of Culebra Creek, flowing out of the flank of the mountains. A long mesa rises sharply to your left, directing the checkerboard flows of water toward the north. Eventually the water reaches the town of San Luis, funnels through a gap in the mesa, and turns south toward the Rio Grande. But Culebra Creek never gains the river. It is exhausted by big Anglo farms west of the mesa.

A few miles up San Francisco Creek is a hamlet alternatively known as San Francisco and La Valley. When Shonnie was a teenager, she rode her horse, Hot Smoke, in the piñons and junipers above the village. San Francisco today consists of some small houses, a little store at an intersection, and a striking, spindly, stuccoed church. A cameo-style portrait of Saint Francis surmounts the door, and a portable confessional—a miniature replica of the church in warped plywood—rests on skids near the steps. As at the Santo Niño cemetery, the landscaping is bare and impoverished. Actually, there *is* no landscaping, the slats sag in the belfry bays, and the building could use some fresh white paint. The overwhelming impression is of a rickety and attenuated faith. But if you walk around and view the church from the side, it's much more substantial, and when you peek in the window, the polished space is full of light.

The church is paired with another building, its spiritual complement, located a little higher on the road: the *morada*. This low, shuttered, brown structure looks as if it wants to worm into the hillside. Part clubhouse and part haunted house, the *morada* hosts the local council of *penitentes*, who are religious devotees. The membership today is down to eight or ten aging men. A century ago, the brotherhood, a force in every village, was banned because of the zeal of the penances, which included self-flagellation. The *penitentes* represented all that was held to be wrong with Spain's Catholic fervor when it was transplanted to America. A little way above the *morada* is a tilted, weathered cross, still used for reenacting Golgotha, and above the cross are the wild woods of the mountains, where witches once roamed.

Back on the straight road to San Luis, it's not far to Marianne and Joseph Medina's restaurant. The two-story building is surrounded by hay fields. It is cupped within a fine prospect of 14,000-foot Mount Blanca, the cynosure of San Luis Valley. Technically a massif or conglomeration of peaks, Blanca steps off the axis of the Sangre de Cristo range like a well-formed pedestrian hailing a cab. The Navajo knew the mountain as *Sisnaajinii* (White Shell Mountain), and it was sacred to them because the first man and woman were said to have emerged from the earth here. The native people likened the features of the landscape to body parts. If the two mountain ranges clasping the Valley are comparable to a rib cage, the breasts are Blanca, the belly is the Taos Plateau, and the thighs are the Rio Grande Valley, fringed on the river banks by willow and cottonwood.

The Medinas' restaurant is called T-ana's. That was Marianne's childhood nickname, an Indian transliteration of Marianne, the T (or *Tae*) meaning Mary. The building is made of adobe bricks and layered with stucco. The color both inside and out is southwestern pink, a muted pink, a reserved pink dreaming of mauve. The couple built the restaurant themselves, stop-and-go. A barterer like his Hispano forebears, Joseph traded logs or did engine repair in exchange for materials or for specialty work like plumbing. Shonnie had been against the project, arguing that her mother's catering business ought to be sufficient and why did they have to expand? Married by then and living in Alamosa, Shonnie had wanted Marianne to spend more time with her on religious service instead. The young woman often got her way but not on this. Construction began in 1997, stopped when Shonnie got sick, and was completed five years after her death. Blanca on this heavy, late Saturday afternoon looked smudged, as if the mountain were painted on the horizon with ashes.

This would be my second meeting with them. The first time, sitting with her husband in a booth at the restaurant, Marianne Medina got right to the heart of the matter. Shonnie had genetic cancer, Marianne said, by which she meant hereditary. He [the doctor] cut her ... The lump was too dense ... Shonnie fought them. ... Marianne's words spilled out defiantly. It spread ... The cancer was always a step ahead of us. She held a piece of paper with her daughter's genetic test result: BRCA1.185delAG. We raised them [Shonnie and her sister] healthy, Marianne insisted, signaling the discord between nature and nurture, between biology and culture.

Marianne this time was in the kitchen, wearing a full white apron, her black hair tucked into a baseball cap. Joseph and Iona, the remaining daughter, were preparing the tables and getting ready to open for the evening. The way the restaurant is laid out, the main dining room is on one side and the booths and a sandwich counter are on the other. In the corridor between the eating areas is a large print by the regional painter Maija, who does scenes featuring Indians and wild animals and raven-haired females. In the Medinas' picture, a young woman, reclining on a bearskin rug, a blazing fire behind her, has come up quizzically onto an elbow; her pale shoulders are bare. A yellow-eyed wolf-dog guards her. Without stretching, you could believe that Shonnie, Iona, or a younger Marianne had modeled for the painting. Superficially the two girls took after their mother, having in common black hair, snapping eyes, and high, flaring nostrils. They were like a team of ebony-maned horses. In fact, Shonnie did model occasionally, just for fun, and Marianne had too, when she was younger.

Genetically speaking, nearly all of Shonnie was contained in the two women and the man at the restaurant. Each person had about half of her DNA. Indeed by an energetic concentration on their bodies a hologram of Shonnie could be willed into being—an extraction of Shonnie, derived from the physical features (the outward form or phenotype) and the hidden DNA (the interior program or genotype) of her three closest relatives. More or less this is how nature had created her in the first place, by making methodical yet unpredictable selections from the chromosomes of her parents.

According to Iona, the sisters didn't look alike. Looks—I got my Mom's, Iona said. Shonnie, she would be told she looked more like Dad. The older sibling of the family was the known beauty, yet some people found Iona even better-looking than Shonnie. Iona herself didn't agree, because if she had gotten her mother's looks, she had also inherited her father's diffident temperament. Iona smiled. 'Heinz 57' I call ourselves, she said. For all the ingredients, you know. We're three or four things mixed. I have more Indian than Shonnie. And she was more Oriental-looking, yes. She had a pointy nose and uplifted eyes. Mine turn down. She had lighter hair, and her skin was porcelain-white.

In Old New Mexico, which was never a melting pot in the democratic sense, genes met and mixed by cruel chance. Later, light skin was the phenotype on which to construct a Spanish pedigree and paper over the Indian one. Neither Iona nor her mother had any problem talking about their blended ancestry. Not quite sure of her own paternity, Marianne thought she might have Chinese blood on top of the rest—hence the slightly Asian cast to her older daughter—though you wouldn't have remarked on any Chinese in Marianne herself. Native Americans having descended from Asians, the younger race could easily have masked the older race in Marianne. A population geneticist wouldn't put it that way.

If Shonnie had such a nice complexion, why was she obsessive about tanning? Year-round tans, Iona explained. That was the look for being noticed. To get it, we would go to tanning booths together or lay out in the sun. We did everything together, Iona said with a level gaze. We were best friends. You plan on having your best friend there with you always. I don't have a best friend even today.

On Saturday nights, if there was a decent crowd, Joseph sang karaoke. A decent crowd for T-ana's was twenty people, but there weren't that many when he began to set up the microphone and karaoke machine at about 6:30. West light slanted into the dining room. One of the long tables was celebrating the birthday of a family member;

two couples sitting at the other tables shifted their chairs to face the low platform where the music was. For a short time in his younger days, Joseph had left the Valley and worked in the musician business, as he put it. He'd played bass in a rock band in Denver and Colorado Springs, and he had a smooth baritone.

Now, at fifty-six, humble, devout, and close to the land, Joseph Medina was a fixture of Culebra. In his attachment to tradition he might well be the *last* Culebran. He was a private, self-reliant man who had made the women of his family the pillars of his life. Most people, even his wife, called him Joe because that was what his father had been called, and though not objecting to Joe, the eldest son and namesake was more formal and fastidious than his father had been. He had fair skin, brown eyes, a small mustache, and a deliberate, dapper way about him that, tonight at least, recalled Xavier Cugat, the band leader of the forties and fifties. Apart from these surface traits, Joseph was what geneticists call an obligate carrier, because, although his DNA was never tested, he had almost certainly transmitted the 185delAG mutation to his daughter. Shonnie's mutation must have come from one parent or the other, and Joseph's family's history pointed to him. He never spoke of it, but he knew.

Once in my life, Joseph crooned, closing his eyes. The song was "Solamente Una Vez," made popular by Cugat and Bing Crosby many years ago. Moving to the middle of the floor, he adjusted the long cord of the microphone. He sang, "Am I blue? Yes, I'm blue," the old Billie Holiday tune, and "Smoke Gets in Your Eyes." The music sashayed from the speakers as the lyrics scrolled on the video monitor, words and sound having been digitized to play together exactly, time after time, just as DNA's digital language, ATGC, the body's tetragrammaton, intones the same message over and over across the generations.

With night on the way, Culebra Peak and the rest of *la sierra* started to glow in anticipation. In the north window of the room, facing Mount Blanca, hovered a green neon sign whose four letters said Open. By now Joseph was glowing too. When he expressed his feelings through his singing, a smoother and more confident personality emerged. This song goes out to my beautiful wife, he announced to the diners, flicking the mike cord professionally. Now don't go throwing anything sharp at me. Then he launched into "You're My Everything."

Wearing an apron, for they alternated as waiters and singers, Iona took the microphone from her father. At thirty-four, she had graceful, broad shoulders and a youthful form, fuller than her sister's. Shiny hair, thick lashes, and large hoop earrings brought out the Indianness of her face. I'll sing a little Patsy Cline for you, Iona said. It was "Walkin' after Midnight," performed in a pretty voice. I'll do "Crazy," she said next. It's your favorite, or it is until you hear me. Which was Iona selling herself short, not like Shonnie if she had been singing. Shonnie's golden rays always swiveled toward the light.

The atmosphere in the dining room was bronze. Joseph squinted at the screen and sweetly sang the Righteous Brothers. "Are you still *mine?*" . . . followed by "When a Man Loves a Woman," his face per- haps too flushed for someone who never took a drink. Then a medley of Mexican songs, which leaned heavily on four basic chords and lurched forward in three-quarter time.

The sun had gone from the room. Putting on his glasses, Joseph sang with the cord fully extended, facing the karaoke machine instead of his few remaining customers. He sang pretty much for himself, comforting himself.

At nine Iona switched off the Open sign. Marianne came out unglam- orously from the kitchen in her apron and ballcap, and sat down at an empty table. Eight years had passed, and all three missed her terribly.

Chapter 2

PREDESTINATION

Joseph Medina's grandfather on his mother's side was named Luis Martinez. Born in the early 1890s—the exact date is not given—Luis grew up in San Francisco and the neighboring Culebra hamlet of San Pablo.

In 1914, Luis Martinez married sixteen-year-old Andrellita Medina of nearby San Luis. Immediately there are grounds for confusion because you see that the name Medina appears on both sides of Joseph's family tree. Crisscrossing lineages are the rule in northern New Mexico and San Luis Valley. This may explain New Mexicans' interest in genealogy—more than a hobby, it's a way to keep things straight before proceeding to the altar. Maria Clara Martinez, who lives in San Luis and who put together a pedigree of Luis Martinez's descendants, has seventy thousand names in her computer, reaching back to the founding of the colony and beyond. Although many Hispanos are able to link themselves to forebears in Spain, usually by way of a conquistador and other notable figures of the New Mexico colony, Clara Martinez can do this for whole communities. It doesn't take many generations for her to prove that almost everyone in Culebra is related by blood.

But to get back to the young marrieds, Luis and Andrellita. Their first child arrived quickly, in January 1915. The baby, Pedro, did not live long, and the couple would lose other children as the years scrolled by. In those days, when infants perished, especially if they died before they were christened, they tended not to make it into the church records that are the mainstay of genealogists. The Martinez descendants don't agree on the number of children the couple had—nineteen is the highest number you hear, which includes a set of twins who died young. Dorothy Martinez, Joseph Medina's mother, was born in 1932 and is still living. Dorothy is at her son's restaurant today, sitting in the front row of chairs in a red dress and white shawl, bright as a button, surrounded by many relatives. She's the oldest of the clan present.

Joseph Medina has a black-and-white photograph of Luis and Andrellita Martinez. The picture, of the couple in profile, delineates the arching Castilian nose and brow of his handsome grandfather and the more rounded features of his grandmother. In the old days a Castilian profile (think Javier Bardem) was a fine thing to have; Andrellita's face seems more *mestiza*. Discussing the photo, Joseph remarked that his grandfather came from Castile. This couldn't be, since Luis's father, according to the family pedigree, was born and baptized in northern New Mexico in 1867. The *pobladores* (little landowners or settlers) of New Mexico didn't flit back and forth to Albuquerque, let alone Europe. Still, you do sometimes hear Hispano people say that their nineteenth- or early twentieth-century relatives emigrated directly from Spain. That is what I have been told, they will say with a shrug. Perhaps they have a vestige of the old *españole* snobbery or maybe they say from Spain simply as shorthand for Spanish ancestry. But all told, pure heritage doesn't matter anymore to the modest people of Culebra, just as claims of *limpieza de sangre* (pure blood that was free of Moorish, Jewish, African, or Native American taint) are long gone from the ex-Kingdom of New Mexico. Nobody would raise an eyebrow if you pointed out that Medina originally was an Arab name.

The reason that family members have gathered at T-ana's Restaurant this Sunday afternoon in 2007 is to learn about their genetic, not cultural, heritage. A genetic counselor named Jeffrey Shaw is coming down from Colorado Springs to speak to the group. Relatives have returned to Culebra from cities north and south because of the breast and ovarian cancer running in the family since at least Andrellita's generation. Our inherited imperfection, Shonnie might have said.

The session with the genetic counselor was a long time in the making. After Shonnie died, Marianne Medina, who is something of a genealogist herself, would take out the piece of paper with her daughter's DNA test result. Brooding over the finding, BRCA1.185delAG, she made a mental list of the cancers she'd heard about among her husband's female relatives. About the same time but acting independently, two of Joseph's cousins in the Denver area started to collect health histories from their parents, uncles and aunts, and other relatives. When the cancer records were superimposed on the Medina-Martinez family tree, the picture was terrifying.

The woman at the top, Andrellita Medina Martinez, the sixteen-year-old bride, developed breast cancer late in life. A midwife cut out the lump, according to one story; another account says that she had a mastectomy and radiation. Andrellita survived and died in her seventies of other causes.

Fourteen of Andrellita's children grew to be adults. The second generation consisted of six men and eight women. Of the eight daughters, five had died of breast or ovarian cancer by the time of the 2007 meeting. Their names were Maria Casilda, Eduvigen (Duvie), Susan, Amalia (Molly), and Mary. Another daughter, ninety-two-year-old Bernarda (Auntie Bernie), was a breast-cancer survivor. On the other side of the ledger, Abel and Salamon, two of the sons, had developed prostate cancer, and Abel had died of the disease. To summarize, half of the Martinez children had had breast, ovarian, or prostate tumors. These cancers are linked in having a hormonal component. The number of cases and the similarity of the pathway screamed out that genes were involved. (It probably was not relevant that Luis, he of the Castilian profile, died of cancer of the tongue or throat.)

Just two of the daughters of Luis and Andrellita were cancer-free:

Dorothy and her younger sister Teresa (Teresita), who was seventy-three. Both were attending today's meeting. None of the Martinez brothers came. When Marianne distributed the invitations, a couple of her male in-laws indicated that they were uncomfortable with the proceedings. They're in denial, said Marianne, not one to equivocate.

Now to the next generation. Exponentially more Martinezes occupy the third tier of the family tree. Dorothy Martinez alone had fourteen babies following her marriage to Joe U. Medina. Eleven survived infancy, including Joseph, the second-born. Although none of Dorothy's offspring had been diagnosed with cancer at the time of the meeting, cancer had already reached around and taken one of her grandchildren, Shonnie. As for Joseph's first cousins and Dorothy's nieces, women in their forties and fifties, some of them here today, they were riddled with breast and ovarian cancer. Eight had had the disease; three were dead. After this generation came an expanding number of offspring who were as yet too young for the disease but who were at risk. Genetic counseling was absolutely in order for this group even without the smoking gun of their DNA, for, in addition to Shonnie, two of Joseph's cousins and one of his sisters had already tested positive for the BRCA1.185delAG mutation. Every member of the clan with a blood connection to Andrellita, the obligate carrier of the mutation, was potentially a carrier.

Genetic counselors are not doctors, but they work alongside doctors. They are health professionals with master's degrees who explain the risks of inherited mutations and the pros and cons of DNA testing to patients and families. Jeffrey Shaw, who worked at the Penrose Cancer Center in Colorado Springs, was an experienced counselor, and this was not his first encounter with the family. He had met Marianne just before Shonnie died, in 1999, and three years later he had counseled Marianne, Iona, and two of their female relatives. Iona was supposed to be tested for the mutation then, but she put it off. I can't pinpoint if I do want to know or don't want to know, Iona said. What good will it do? I go every other year for a physical breast exam. I'm OK now, and if I get it, I get it. Another time Iona said, Call me crazy, but I still don't want to know what may be lurking in my body. Like many women in her situation, she imagined the gene, if she carried it, as a tiny bit of cancer in itself, so small that it might be repressed into nothingness. Today, as Jeff Shaw drove down from the Springs and through the mountain wall to Culebra, he thought to himself, She is too sweet to ignore this. On entering T-ana's, he jocosely scolded Iona, saying to her, You are on my *shit* list.

Goateed and a little chunky, with a breezy, affable manner, Shaw wore shorts and a checked shirt to the weekend session. He set up an easel and poster paper at one end of the dining room, where Joseph and Iona had performed the night before. About three dozen family members occupied the tables and booths and rows of extra chairs. Some of the parents had brought their children. Waiting for the event to begin, a teenager trained her video camera on her cousins. Their collars too big and their hair plastered down, the boys scowled back as boys do.

Joseph, the grave host in sunglasses and a charcoal blue suit, stood in a back corner. During Shaw's talk he moved about restlessly. With silent reservations he had gone along with his wife's wishes that this health matter of the family should be brought into the open. But what did he think about taking the gene test himself? It'd be OK, he said, clearly in no hurry. Marianne, having prepared the chips and dip, the lemonade and soft drinks, sat with a bunch of her in-laws. Wearing a dark dress, she looked at ease. Iona, her black hair pulled back with a silver clasp, minded the refreshments, lowering her head as she went in and out of the kitchen. Married not long before Shonnie fell ill, Iona and her husband had forsworn having children—not because of the cancer risk but because of the fraught state of the world, a world they believed was coming to an end.

The afternoon was warm, and soon the dining room was warm too, creating a run on the lemonade. Joseph's younger sister Louisa arrived in tight red shorts, definitely the least Hispano of the Medinas. Skittish Louisa had decided that if she were going to have the gene test, she wouldn't want to know the result, although later she would change her mind. Another sister, Wanda, flashed a friendly smile but looked hollow-eyed and drained from her harsh, therapeutic diet. She believed she already was fighting the disease. A blonde cousin, Linda, was here because her sister Gean had died recently of breast cancer. Shaking her head, Marianne said that Gean had chosen to have her tumor treated with surgery and chemotherapy. She had a biopsy and it went crazy, Marianne said, and *still* she did conventional medicine.

Joseph's niece Shannon, representing the fourth generation of the family, came with her thirteen-year-old son. Shannon's roots were in Culebra, but she, like most of the family, had moved on and branched out. In addition to the names Martinez, Medina, Chavez, Sanchez-Vigil, and Salcido, there were now Kahn-Ortiz, Kramer, Pakalenka, Monarski, Hosack, and Rich-Crane. For the first time in centuries the blood of the Martinezes flowed outward rather than inward, seeping into other landscapes, mingling with other streams. With dilution, a certain genetic disinfection takes place in the American melting pot, especially as families become smaller and better informed about heritable diseases.

Jeff Shaw spoke to the group that day about the BRCA1 gene. BRCA stands for breast cancer, and in the mid-1990s BRCA1 was the first such gene to be identified, hence the name, followed closely by the discovery of BRCA2. The disease caused by the two genes is called HBOC, for heritable breast and ovarian cancer. A woman who carries a mutation of the genes may contract one or the other or occasionally both of the conditions.

BRCA1 is not a breast-cancer gene, Shaw said, right off addressing a pervasive misunderstanding. It is a tumor-suppressor gene, which everyone has. The misunderstanding was this: In the nomenclature of medical genetics, the tail wags the dog. A BRCA1 mutation is a rare, flawed variant of the normal gene. The gene was given the name of its mutation after the mutation was discovered. There is some logic to the process because if it hadn't been for the mutation (the anomaly), the normal type would not have come to light. By the same token, the function of the gene—to keep a cell in good repair, to keep tumors from occurring—wouldn't have been discovered if the failure of the gene hadn't manifested itself first. Like a broken pane of glass, the mutation exposed a window into the body that no one had noticed before. Be that as it may, scientists discussing BRCA1 toggle back and forth; from the context, they know which state of the gene is meant, normal or mutated.

Shaw skipped the fine points, which were more than the attentive Martinezes and Medinas needed to hear. Taking his felt-tip pen to the poster paper, he sketched a squiggly pair of chromosomes—in each human cell, chromosomes come in pairs—and he marked a point on one of the chromosomes where the BRCA1 gene resided. Likewise for the BRCA1 gene on the opposite chromosome. The two genes have separate sources, he observed. Two copies of the gene are passed to you, one from each parent. Joseph, in the back, shifted uneasily.

Then Shaw started to explain what happens when a mutation occurs inside a cell. I prefer to call it a boo-boo, he said to the group. Granted, mutation was a distasteful word, which Shaw was not alone in disliking. Marker, another term he used, was an improvement but not exact. A mutation simply means a change, a change in the biochemical spelling of a gene, and not all such changes are harmful. This mutation surely was, however.

The theory is, it takes two hits on the DNA in order for a cell to turn cancerous: two hits meaning that both copies of BRCA1 are disabled. A hit can be inherited, as when a child receives a BRCA1 mutation from a parent. Then they become cells with no backup, Shaw said, making an X on the chromosome he'd drawn. At a later time in life, a second hit occurs—say, from radiation or a toxic chemical or just an accident of nature, and it disables the second copy of BRCA1 in a breast cell. Two more rapid squeaks of the pen. Now *both* of the tumor-suppressor genes are damaged and they lose control, he said. The rogue cell divides and multiplies until there is an uncontrolled growth of cells.

The two-hit theory of cancer explains the difference between the breast cancer that most patients get and HBOC, the condition plaguing the Medina-Martinezes. Most breast cancer is called sporadic, a term indicating that it has no familial pattern. Sporadic cancers cannot be assigned to the hard-wired inheritance that was plain in the pedigree of the people here. That notorious number that a woman hears, one in eight, describing her risk over a lifetime, is based on the sporadic cases of breast cancer. Nine out of ten cases of breast cancer are sporadic, their causes unclear. Genes no doubt are involved, but they can't be very powerful genes, or else scientists would have spotted them by now.

When a breast cell first takes a wrong turn, tumor-suppressor genes like BRCA1 and BRCA2 step in and try to fix it. You can appreciate that if a woman begins her life with two working copies of an anticancer gene, and one gets knocked out along the way, she's still covered. The second hit on the repair gene might never happen, or, if it does, the odds are that it will strike later in her life, which is why sporadic breast cancer is generally a disease of older women. In that regard, you could say that Shonnie Medina began her life with one hand tied behind her back. Contrary to all appearances, she had a physical handicap.

Inheritors of a BRCA1 mutation aren't absolutely guaranteed to get breast cancer. A range of risk is given: The high-end estimate for carriers in families like the Medinas and Martinezes is more than 80 percent, that is, an 80 percent chance over a woman's lifetime, while other studies (fewer) have found that the risk may be as low as 35 percent. The risk probabilities for ovarian cancer are lower, in the range of 40 percent, which is still significant. Male breast cancer, normally a very rare condition, takes an upward tick in male BRCA carriers. But with female breast cancer, what distinguishes the heritable cases from the sporadic sort is the skew toward younger women.

There is a 50 percent chance that if it happens, Shaw said to the group, it will happen before the age of fifty. A family history of breast cancer should be a red flag, he said, and it's a red flag if a doctor sees a young woman from that family who's concerned about a suspicious lump in her breast. Shaw obviously was thinking of Shonnie when he quoted the hypothetical physician saying to his patient, Oh, you're only twenty-four or twenty-five. That has to be a cyst. Go home, get out of here. . . . No! Shaw said. We need to know if you've got cancer in the family.

All this was going down smoothly at T-ana's. A couple of times, a cell phone rang and was quickly turned off. Now the counselor mounted a very effective demonstration of the inheritance pattern of BRCA1. Borrowing from the refreshments, he used a red Coke can to represent the mutation and three green cans of Sierra Mist to stand for copies of the normal gene. He called up Debbie Rich-Crane, Teresita's daughter and Joseph's cousin, to pretend to be his wife. Debbie was an imposing woman about Jeff Shaw's age. After some ice-breaking banter, he gave her two green cans to hold while he held out a red can and a green one. When he said, Our child will inherit one can from each person, the audience readily grasped the fifty-fifty probability that the red can, the adverse mutation, would be passed along.

Importantly, Shaw had made himself the carrier, the male who masks the breast- and ovarian-cancer risk and transmits it to the next generation. It doesn't only pass from mothers to daughters, as when Debbie held the red can. The cancer can skip generations, Shaw said, especially if it goes through men. Most doctors will say your dad's side doesn't matter. No. Then Jeff and Debbie exchanged cans. Whatever combination of BRCA1 genes they presented, always there was a 50 percent risk that their child would turn out to be a carrier of the mutation. The counselor added a warning, mindful of the large families present: Remember, a coin can turn up tails four times in a row.

What if the baby inherits *two* copies of the mutation? someone asked. That does not happen, Shaw said. A miscarriage would occur and the baby would not make it. Shaw stepped over a permutation that might well apply to the people in this room, however. If two BRCA1 carriers, young and poor and not knowing their risk, met and married, they'd stand an even greater chance of producing carriers in the next generation, while their miscarriages would not be remarked upon. Given the crossed bloodlines in the community, it was not out of the question that *both* Luis and Andrellita had carried the mutation.

. . . and now she is sixty-five years old, Shaw was saying to another questioner, and she does not have cancer. Is it possible for *her* to be a carrier? Yes! Shaw could have been speaking about Joseph's mother, Dorothy. Though disease-free, she, like her oldest son, was an obligate carrier, as in has-to-be, since there was no other way to explain the path of cancer in her family. Dorothy Medina was paying attention to Shaw's talk mainly out of politeness. Her English wasn't the best. Urged by her daughters, she did go for DNA testing a couple of weeks later and got her positive result. Joseph, impassive behind dark glasses, never took the test.

A cousin named Beverly Ortiz raised her hand and asked about screening. Beverly, who lived near Albuquerque, was one of the few who already knew she was BRCA1-positive. She had taken the test because her mother had died of ovarian cancer and her sister had contracted breast cancer. After her childbearing years, Beverly had her ovaries and uterus removed, and today she kept a close watch on her breast health through mammograms and MRI screening. In the eyes of the medical profession Beverly Ortiz was an ideal patient, informed and proactive. Her effort to outflank her genes so far had succeeded. Shaw said to her approvingly, For positives, yes, we're going to do increased screenings. And there is a huge drop in risk by doing the surgical intervention—by which Shaw meant the ovarian removal. He seemed to recognize that this crowd wouldn't countenance preventive mastectomy.

Shaw gestured toward a stack of pink handouts on the table. Here's the info. You decide. It has my phone number and my e-mail address. I will get back to you promptly if you contact me, but I'm a lot better with e-mail than with the phone. Although genetic counselors are trained not to twist patients' arms about DNA testing, when you make an appointment with them, they make their preferences ringingly clear. Along with nurse practitioners, physician's assistants, and other clinical staff short of MDs, they buy into the rationalism of scientific medicine, yet they haven't become hard of hearing of emotions.

The questioning meandered off-topic as the session wound down. Fidgety and flushed with lemonade, the kids acted as litmus strips for the moods of their parents. Debbie Rich-Crane wanted to know Shaw's opinion of homeopathic therapy for cancer. Though firmly against it, the counselor tiptoed around the subject, having been clued in that it was a Medina issue. I don't know much about it, he said. Exercise can reduce the risks . . . It's not my area.

How did this family get the gene? At long last—the crucial question. A few members of the group already knew the story of the mutation's ancestral source. They'd been told by Debbie or another cousin who had come across the story while investigating the family cancers. Indeed, at least one of Joseph's brothers had stayed away from today's meeting because he refused to accept the story that the Martinezes and Medinas, Spanish on the surface, with some Native American below, were also descended from Jews.

Yes, Jews. You hear all the time that America is a melting pot, but if so, the mix is lumpy and the ingredients may not be as stated in the family recipes. For the Hispano Catholics of northern New Mexico and southern Colorado, Jewish ancestry was a will-o'-the-wisp of memory and culture, which many people had heard about without knowing if it was true. Again: What was a Jewish gene doing in people who maintained they were Spanish?

Jeffrey Shaw hesitated. Several years before, Shaw himself had helped to uncover the surprising fact. Essentially, the DNA that he and his colleagues found in the San Luis Valley confirmed events that had happened centuries earlier in Spain and that were echoing still. Perhaps the genetic counselor didn't care to stray into a social minefield. Researchers have been told to get out of Hispano households when they mentioned Jewish genes. In any event, Shaw stopped speaking and turned around, looking for assistance. How did this family get the gene?

That was my call to speak. The story I gave was a digest of information that will be given in later chapters of the book. Most likely the mutation arrived by way of Sephardic Jews who converted to Catholicism half a millennium ago under pressure from the Spanish Inquisition. The Martinez descendants listened, not exactly blankly, their faces incorporating the story of their Jewish forebears for further review. As stucco is smoothed over adobe, people may layer an identity for themselves at odds with what lies beneath, and no longer recall having done so.

Taking over again, Shaw fleshed out the specifics of the family mutation. He explained that BRCA1 was a very long gene, thousands of biochemical letters long, and therefore it had thousands of places where misspellings could occur. So at letter number 185, he said, there's been a deletion of the A and the G; A, for adenine, and G, for guanine, are two of the four letters of the DNA code. That's why the boo-boo is called 185delAG, he said. When we test members of this family, Myriad [Myriad Genetics, the company with a patent on the gene test] will look at only that one place in the gene.

When Shaw then characterized 185delAG as a *Jewish* marker, his voice quite properly put quotes around the phrase, for Jewish marker or Jewish gene is misleading. Jewish *ancestral* marker is a better way to put it because carriers of the mutation may also have ancestors who were *not* Jews, clearly the case here.

Shaw then made his good-byes, mingling for a few minutes, and took off for Colorado Springs. That was a shame, because he missed the afterpiece, wherein the members of the Martinez and Medina clan dropped their seriousness and celebrated their family reunion. People produced digital cameras and took pictures of one another, to much oohing and aahing. The room filled with happy noise. Dorothy Martinez, the smiling senior member, looked somewhat like a party pretzel with her bent back, sticklike limbs, and salt-colored hair. Requests were called out for Joseph to sing, and while he obligingly got the equipment ready, a master of ceremonies materialized.

Bill Kramer was Joseph's brother-in-law. He and his wife, Wanda, lived in Alamosa, the Anglo hub of the Valley. Pale, with a shaved head and dancing, glacier-blue eyes, Bill poked fun at himself as the so-called German outlaw of the family. For employment, he was a quiltmaker and a snow-shoveler, among other jobs, while Wanda earned money cleaning houses. The answering machine where they lived was maxed out with unculled messages. Joseph Medina, as the head of the family, was not unfriendly to Bill, but he was puzzled by his artistic tastes and offbeat ideas. In another era, Bill might have been a troubadour, or a holy fool running after Saint Francis, or a gypsy. He had a big, overflowing heart.

So Bill commanded the floor at T-ana's. He told an elaborate joke about hunters in the woods, acting out two or three of the parts and almost pinching himself in delight. He didn't want any of his in-laws to go home early. After Joseph started singing—faster music than the night before—Bill jumped up and began line-dancing with his wife and sisters-in-law and some of the female cousins. "Boot-Scootin' Boogie," yeah, came pulsing from the karaoke speakers. Line-dancing was a contemporary Western form belonging more to Alamosa than Culebra, but the exuberance on display was all Hispano. Joseph's sisters Lupita and Chavela especially could twirl. Between songs Bill had to admit, with a whoop of laughter, that it sure didn't look like a Jewish dance. Taking the mike himself, he sang, "Who Put the Bomp (in the Bomp, Bomp, Bomp)," and it went surprisingly well.

If Shonnie had been here, she would have been right at the center of the singing and dancing. To deflect their admiration of her, sooner or later people would have started to tease her or tell stories about her. The blonde, the airhead, they called her, because she was klutzy as well as beautiful. How she always got her fishing line tangled. Not paying attention and running out of gas. How about the time the blanket slipped off the roof of the car and got caught in the axle, and Shonnie driving along without a clue until the car stopped dead in the middle of the road? MICHAEL, she would wail—that heedless, histrionic wail of hers—summoning her husband to the rescue. If only Shonnie had been here.

Wanda Kramer, Shonnie's aunt, looked better than she had at the beginning of the afternoon. A warm woman with cascading black ringlets of hair, Wanda was convinced that she had a breast tumor. Some weeks earlier, a radiologist had spotted a lesion on her ultrasound scan. Alarmed, the radiologist had urged additional tests. Bill Kramer and the radiologist had butted heads over the necessity for a biopsy. Another issue was cost, since the couple didn't have health insurance.

Bill related all this privately, as the party at T-ana's ebbed. The good news was that the lesion in Wanda's breast had started to shrink, Bill said, thanks to her stringent diet and other measures of natural healing. A student and part-time salesman of alternative medical approaches, Bill spoke confidently about the body's response to synergistic blends of essential oils. I believe in mind–body, he said, his brow glistening from the dancing. That's my paradigm. I've seen so many amazing stories. I'm waiting for the doctors to become more open-minded. And if we're going to do natural healing anyway, why does a biopsy matter? A biopsy would only cause it to spread, Bill concluded sunnily. If only the strategy had worked out.

He talked about Shonnie, of whom he'd been very fond. While Shonnie was sick, Bill had the idea to raise money for her care from sales of his handmade quilts. He set up a table outside the supermarket and made a sign saying, I am raising money to help my niece who has breast cancer. He was able to give her $640 from the proceeds.

An emotional man, Bill teared up. Shonnie always saw the good in everyone, he said. She told me—I remember how it cheered me up—she told me, I see an elder in you. Uncle Bill, you will become an elder.

Shonnie believed that Bill was a good enough person to become an elder in the local congregation of Jehovah's Witnesses. For she, Bill, Wanda, Iona, Joseph, Marianne, Dorothy, and a number of other Medinas had converted to the Jehovah's Witnesses in the 1980s, rejecting the Catholic religion they'd known as children. The serene millenarianism of the Jehovah's Witnesses had taken root in the thin air of Culebra, like an epiphyte growing upon a Catholic tree but taking no nourishment from it. Perhaps that's why Bill was so upbeat. The nettlesome questions of life and death and race and strife didn't bother the Witnesses as much as other people, because of the promise of the last days and the Earthly Paradise soon to come.

The Jehovah's Witnesses were at the far end of a revolution that had been in motion in the Western world since the Crucifixion. The Christians had rejected the Jews, the Protestants the Catholics, and then the Witnesses had broken from the Protestants and all who came before or since, each new group maintaining they had returned to the fundamentals of the faith. How plastic are the choices that human beings make, ever spinning off to new landscapes. But if you are a genetic determinist, you wonder about the centripetal pull of DNA.

Chapter 3

THE WANDERING GENE

So God instructed Abraham to go and found a great nation in the land of Canaan. Abraham did as he was told. Abraham was a Jewish knight-errant. After sallying back and forth, he fathered a family and settled in Palestine. This happened (if it happened that way) three or four thousand years ago. Armed by an agreement with their God, the Jewish people came into being. First a family, then a clan, then a tribe, finally they became a family of tribes wresting territory from other tribes. The land they took over was dry, and was watered by dikes and channels in the manner of the Culebran *acequias*.

Politically they were called Hebrews or Israelites or Judeans. Genetically they were blends of the peoples of the region, combining DNA from the original Canaanites, Amorites, Hittites, and others. None of the other Semitic groups lasted, and why didn't they? Their genes live on in today's Jews, but the cultural cladding for their genes sloughed off and fell away. Naturally, peoples and nations rise and fall. Why not the Hebrews then? Living at the busiest intersection of the Mediterranean, they suffered heavy-footed incursions from Egypt, Assyria, Babylon, Persia, Greece, and Rome—much bigger civilizations that did not last either.

And again, whenever the Jews were fragmented and forced from Pal-

37

estine, which happened more than once, still they held together. Along the highways of the ancient world, all sorts of peoples, not just Jews, were constantly being dispersed by famines, epidemics, conquests, and other upheavals. The refugees assimilated where they landed, surrendering their ethnic passports to the locals, as it were, and distributing their genes accordingly. But not the Jews. Or not readily, at any rate. It's as if history attached a locator-beacon to them, a cultural GPS, so that Jewish communities can be tracked around the world and down through time.

As told in the Hebrew Bible, the Israelites' fierce prophets kept them united during their displacement by reminding them of their obligations to God. Having drifted from Canaan to Egypt, they were guided and goaded back to the promised land by Moses. After Jerusalem was destroyed in the sixth century BCE, and the heart of Jewry was deported weeping to Babylon, a hot-eyed patriot named Ezekiel came forward. Ezekiel had visions of bones in the desert coming to life, of Jerusalem restored. The prophet lectured the Babylonian exiles on maintaining their blood purity. He urged them to guard their sacred separateness lest the pagans around them swallow them up. Reiterating God's covenant, Ezekiel said that God would bless the Jewish people and multiply them, and that God would set His sanctuary in the midst of them forever. Consider the image of the sanctuary, physically set in the middle of them. The sanctuary was like the nucleus of their cell. Their backs turned to outsiders, their eyes locked upon the sanctuary, the Jews seemed to some a stiff-necked people.

Ezekiel's successors, Ezra and Nehemiah, hammered upon the hatefulness of intermarriage. The holy seed have mingled themselves with the people of those lands, Ezra inveighed. He demanded that men get rid of their foreign wives. Indeed, the Hebrew Bible's repeated admonitions against taking up with foreign women suggest that the practice was fairly common. Hence also the emphasis in the scriptures on genealogy, on continuity of lineage, on all those tongue-twisting names connected by *begat*. Judaism's bond was like an epoxy, the resin of religion mixed with the hardener of blood, nature interlocking with nurture. But of the two agents, blood was the more telling. Any person with Jewish fore-bears, traditionally through the mother's line, was deemed to be a Jew—that was his or her default identity. What the person actually attested to was secondary, even irrelevant.

So Jewishness had an in-going, biological dimension, and genealogy was the means of affirming it. It is in this period that the idea of a Jewish race starts to take shape, race representing a seamless new construct of blood and culture, more potent than either blood or culture separately. Formerly, the units of encounter between populations were tribal rather than racial, and the typical Middle Eastern emigrant would have found his genes advancing ahead of him, so to speak. The individuals he or she met would be distantly familiar, striking a chord from earlier, unre-membered exchanges of DNA. Racial awareness changed the perspec-tives of both parties. When a proud and self-contained people such as the Jews, riding the high trajectory of their faith, splashed down in a foreign territory, they probably stirred up more than a garden-variety fear of strangers.

But Jerusalem—Jerusalem was their home. When a portion of the Judeans returned to Jerusalem from exile in Babylon, foreigners abounded in their citadel. The high priests felt the national identity needed to be tightened. If DNA analysis had been available, the Hebrew prophets might have become even tougher guardians of the racial purity of Jews. A panel of Jewish genetic markers might have been created. Lining up at the rebuilt Temple, the people would have provided DNA swabbed from their cheeks. The racial test—analyzing the entire genome or inventory of their genetic material—would have allowed for the earlier borrowings from Semitic groups. That is, a certain percentage of Canaanite heritage would be expected, and, within ranges, other foreign genes would be grandfathered into the criteria as well. Conversely, among the diagnostic markers of Jewishness, the new breast-cancer mutation 185delAG might have been flagged because it's nearly surefire proof of Jewish ancestry.

In any event, by looking at a sufficient number of locations in peo-ple's genomes, science would be able to tell the religious authorities who was a Jew and who was not, and men would be advised whom they could marry and not. Tests like this exist today and are starting to be used, and sharp-tongued prophets of genetics are being heard too.

In 2001, Dr. Harry Ostrer, head of the Human Genetics Program at the New York University School of Medicine, published a deeply researched paper in which he declared that Jewishness is not determined exclusively by DNA but that DNA can help to provide dispersed populations of Jews with a group identity. Ostrer thought that, thanks to modern genetics, the issue of whether Jews constituted a race, a people, or a genetic isolate could be confronted head-on.

Asked to elaborate, he e-mailed: The Jewish genome is like a series of Cluny tapestries. Threads have been woven in at different times, but once there, they become recognizable—recognizably Jewish. In our parlance, they become ancestry-informative markers. These threads are influenced by genetic isolation, drift and selection, which makes them prevalent in a large number of Jews.

But I was making the point in my paper, Harry continued, that Jewishness is also a religion, so you, waspy Jeff Wheelwright, could study Talmud with a rabbi, pass a *Beit Din* (rabbinical court), and take a *mikvah* (ritual bath) and be just as Jewish as I am. (Circumcision is a requirement too, if not performed already.)

The Hebrews' insistence on blood purity had biological consequences. One term that Ostrer noted in his e-mail was genetic isolation. It means that because of inward mating practices, the group's DNA was walled off from other populations. Endogamy and consanguinity are related terms he might have used, steering away from that old-fashioned pejorative, inbreeding. But inbreeding and its opposite, outbreeding, are simple, sturdy concepts. Outbreeding is nature's way of keeping harmful genetic mutations from building up within a population. As a rule,

DNA faithfully reproduces its text, handing down identically spelled genes across the generations, but with so many genetic reproductions taking place, mistakes in printing inevitably occur, and a few of the mistakes cause disease. If a harmful mutation can now be compared with a rock flung into a lake, you'll observe that the ripples from the mutation lengthen, suggesting that the altered gene can show up in multiple descendants. Just as important, the ripples weaken as the effect spreads outward. Half of the time a new mutation is not passed down at all. With outbreeding, harmful mutations diffuse and disappear into the swirling lake of humanity.

Now when the same rock is thrown into a small pool, the ripples hit the sides of the pool, slosh back and merge with succeeding ripples (the following generations), magnifying the effect. That's inbreeding. The blood ties of an inbred group overlap, and carriers of mutations who unknowingly are cousins are more likely to meet each other and reproduce. Genetic drift, another mechanism that Harry Ostrer mentioned, evokes the haphazardness of the DNA rippling in the pool. Genetic drift refers to the role of chance in increasing or reducing the frequency of a mutation. Because of drift, a mutation affecting a few families in a small, inbred population may gain an outsize place in the population once that group expands. And the Jews did expand, if only by virtue of their longevity—two dogged steps forward in the world for every painful step back.

Harry's e-mail touched on a third factor at work in the Jewish genome: selection. This is the Darwinian process working at the level of genes. Simply put, natural selection means the bad stuff is weeded out because carriers of harmful mutations usually don't do as well as their rivals in the mating game. But when individuals carry good stuff—genes that respond favorably to the environment and boost the carriers' fitness and possibilities for finding mates—their DNA is likely to multiply in subsequent generations. As new mutations emerge at random in a population, natural selection puts them to the test. Occasionally there will be a change in the DNA that both hurts and helps an individual, as it makes her less fit in one respect but more fit in another. (The classic example is sickle-cell anemia, discussed below.) It's not impossible that 185delAG, a dark cloud of disadvantage in the human genome, has a silver lining too. The mutation might have been beneficial in the past, when environmental conditions were different.

The breast-cancer mutation 185delAG entered the gene pool of Judeans around the time of the Babylonian captivity, some twenty-five hundred years ago. Random and unbidden, it appeared on the chromosome of a single person, who is known as the founder. In the same sense that Abraham is said to have founded the Jewish people, scientists call the person at the top of a genetic pyramid a founder. Whether a man or a woman, this particular founder was born missing the letters A and G at the 185 site on one copy of his or her BRCA1 gene. The deletion of the two letters interrupted the genetic code and disabled the gene's protective function. The mutation wasn't immediately harmful to the founder because he or she had another copy of the gene that worked.

The 185delAG mutation is very old. How do the researchers know when its founder lived? The date was deduced from historical evidence. When Jerusalem was restored to them, not all the Judean exiles returned from Babylon. The ones who stayed behind are the ancestors of Iraqi Jews, who are today much reduced in number, but for centuries they were a venerable center of the faith. In addition to the Jews living in Mesopotamia and Jerusalem, satellite communities had sprung up elsewhere in the Middle East. A decentralization of the gene pool had begun, and the distances between groups acted as barriers to the exchange of DNA, barriers that persisted into the modern era. In the 1990s, when scientists in Israel tested BRCA1 carriers from the dispersed Jewish populations, they discovered that all had the same spelling in the genetic region of 185delAG. Clearly this was a universal Jewish mutation. But some of the matches between groups were off by a letter or two, which indicated minor changes since they had split apart. Rolling back the demographic clock, the scientists inferred that the mutation's founder must have lived before the groups divided—i.e., prior to the Babylonian watershed.

Keep in mind that this sort of research, dating a genetic mutation, could not have been done on another people. It requires the subjects to have maintained historical records—the cultural cladding around their genes—for three millennia. Inbreeding was a helpful factor too, because it created consistent hallmarks within the hodgepodge of DNA. Even Zionism abetted the task, because it led to the state of Israel, which drew Jews from all over the world and recentralized the DNA, thereby easing the research costs. Not to be overlooked are the native interests of the researchers themselves, what sociologists delicately call the ethnic concordance. Since many of the world's geneticists are Jewish, they naturally have been drawn to study themselves.

The understanding is not only that 185delAG originated with Jews but also that when the mutation shows up in another ethnic or racial group, such as Shonnie Medina's Hispanos, it's because a Jew or a descendant of a Jew has married in. Beyond the bounds of Jewry, sightings of 185delAG are few. There are scattershot reports in the medical literature of gentile carriers in Spain, Chile, Slovakia, the Netherlands, Pakistan, India, even South Africa. Most of these countries have or used to have Jewish enclaves. By the same token, there are American carriers who are not Jewish, or so it's reported on their orders for gene tests, but these people, like the Medinas, undoubtedly have Jewish ancestors. As an example, the mutation has affected a large and gregarious family from the Midwest named K———. Irish Catholic on the mother's side, German Protestant on the father's, the family suspects that the gene came in from the German side.

A corollary theory about 185delAG is that it occurred just once in history. However, nothing about DNA prevents lightning from having struck twice at the AG site, altering the BRCA1 gene of two different founders at two different times in history. Although evidence for a twenty-five-hundred-year-old Jewish founder is overwhelming, couldn't there be others whose descendants are not as numerous or as conspicuous? Several claims of a rival 185delAG have been put forward, only to fade when the fine print of the DNA region is analyzed and the signature of the Jewish variant emerges.

One case of independent origin is strong, though. It comes from

Yorkshire, in England. Two families there, not Jewish, share a novel spelling around the site of their AG deletion, which points to a different founder. The geneticists who study BRCA are not all in agreement that the Yorkshire mutation is truly new. Genotyping techniques have advanced in the fifteen years since the mutation was found, and if the experts were bothered enough by the uncertainty, they could undertake a closer analysis of the DNA samples in England, but as of this writing they haven't done so.

Harry Ostrer was in the camp that believed that 185delAG had appeared in a Jewish founder just once in history and had spread from there. Ergo, the Yorkshire carriers must be related to Jews.

Well, I asked, why hadn't Harry and his colleagues straightened the matter out?

It's not rocket science, he replied, and the work wouldn't cost that much. Pressed again on the issue, he said, You should do a *shidduch*, Mr. Jewish Wannabe, and get [the Yorkshire scientists] to send the samples to [the Israeli scientists] for analysis.

So the gene was present in the people at the time of the Babylonian Diaspora. In the five centuries preceding the birth of Christ, the people of Judea resisted the paganizing influences of their successive overlords: Persian, Greek, Syrian, Roman. Indeed, during the latter part of this period they made converts among other peoples, bringing substantial new blood into the fold without relaxing their own strict standards. But their proud state came crashing down, the Second Temple destroyed, after the Judeans rebelled against the Romans in 66 CE. The Romans killed or enslaved everyone they did not drive from Jerusalem. Judaism from that time onward was a religion in flight, never confident, wherever it set down roots, that it would not be ripped out.

There followed a fluid period in Palestine in which pagans, Jews, and

Christians jockeyed under the slackening eye of Rome. The empire, as it weakened, succumbed to Christianity. Now Christianity provides the backdrop for the Jewish Diaspora. Christianizing peoples are the broad canvas on which Jews' tracks can be discerned after the collapse of the Roman Empire. The Jesus movement hadn't set the Jews in motion, but its astonishing success ensured that the Jews would not come to rest. In evolutionary genetics, as opposed to medical genetics, if a new mutation with a bodily function doesn't make its host stronger, it will fade out and be replaced by a change that does. Random and unbidden, the new, Christian variant of Judaism soon eclipsed the ancestral type, taking over Western Europe, North Africa, and Palestine.

Until there were Christians, Jewish chosenness was not a problem. That the Jews were the chosen people of God was something they told themselves as part of their rubric of unity. The paradoxical responsibility of an insular people was to shine the light of their chosenness on all of mankind, yet this was a duty they must perform indirectly, by obeying the laws of the covenant. Chosenness was their own affair, really, until Christians objected and, flinging *Christ-killer* into the faces of Jews, appropriated the idea of the elect for their own believers. Ah, well. The Jews were prepared to be disliked, persecuted, exiled. Their Bible had warned them. Exile and suffering were the price of chosenness. If God intended them to be a beacon apart from humanity while lighting the way for humanity, they should not wonder that people would throw stones at them from the darkness.

Islam was born in the seventh century and conquered Palestine, ending any hope that the Jews might be reinstated in their homeland. For a while, Muslim and Jewish clans lived together in peace in Medina, which was Muhammad's capital city in what is now Saudi Arabia, but the peace broke down. The Jews wandered about the Mediterranean and crossed the Alps. They waxed and waned according to the tolerance of the societies in which they were embedded. Certainly tens and perhaps hundreds of thousands of Jews lived in Rome, in southern France, and along the Rhine, while a separate group migrated to the Iberian Peninsula. Carriers of the185delAG breast-cancer mutation must have been among them, but of course there are no records of the gene's toll. Life spans were so brief that a Jewish wife would bear her children and die of other causes before fast-acting 185delAG could catch up with her.

A curving corridor from Italy to France to Germany is thought to have brought bands of Jews to the northern region they called Ashkenaz by the year 1000 CE. Many men would have brought along wives and children, but just as many, the adventuring males, traveled solo and would take local mates. Some of the genetic data suggests that male pioneers of Ashkenaz took foreign wives and literally changed them into Jews. When DNA from different ethnic or racial groups comes together, the technical term for the result is admixture. But the Jews' admixture or outbreeding with Germanic peoples did not last long. After they established their settlements and families, they circled the wagons and resumed their endogamous practices, their rabbis hectoring them about blood purity in a new Germanic language, Yiddish. Thus did the Ashkenazim come into being. They are by far the world's most numerous Jews today. Almost six million Americans have Ashkenazi ancestry; they represent about 90 percent of American Jews.

Harry Ostrer and other Ashkenazi scientists debated the genetic ratio between the original Jewish migrants from the Middle East and the Europeans they eventually became. Historically, there has been a high-admixture crowd and a low-admixture crowd, Harry explained. Evidence of high admixture justified Jews' rightful assimilation into Western society, since they were difficult to distinguish from other Europeans. Whereas findings of low genetic admixture supported their ethnic pride: the idea that Jews were a people from Palestine, as told in their Bible.

A persistent third faction argued that the Ashkenazim were not from the Mediterranean region but rather descended from Turkic converts, the Khazars, who migrated to Europe from western Asia. The DNA evidence for the Khazar theory of origin was rather weak, however, and was dispatched for good by a paper that Ostrer and his associates published in the *American Journal of Human Genetics* in 2010. The latest research reaffirmed Jews' roots in the Levant.

But the Khazar theory has the advantage, Harry observed, of absolving the majority of Jews from Christ-killing.

Do you mean that some Jews could say, Well, at the time of the Crucifixion my ancestors were out of the region—don't blame me?

Yes, he replied, because the Khazars were not talking to Pontius Pilate.

After the year 1000, when Christian authorities in Europe banned intermarriage with Jews, isolation became a two-way street. Prohibited from farming and landowning, Jews fastened themselves to urban centers and found success in trade, medicine, and moneylending. The latter service was especially useful to Christians because the Catholic Church did not permit its own people to charge interest on loans. During this period, the Jews burnished the Talmud while Christians burnished the myth of the Wandering Jew, an ageless sinner who could not die until the promised Second Coming of Christ. A corrosive slander against Jews was that they were required to drink the blood of Christian children.

Now come the Ashkenazi bottlenecks. Classically, in a bottleneck a population sharply contracts and then just as sharply expands. The passage through the bottleneck shears away some of the DNA—that is, it reduces the number of variants of a group's genes. The remaining variants arrange themselves in different proportions. Mutations too scarce to make a dent on a larger population can be enriched. Thus the likes of 185delAG might come through a bottleneck not only unscathed but more prominent than before, its carriers overrepresented in the new population. Genetic drift, as noted earlier, succeeds a bottleneck. Genetic drift can happen when a people recovers from an abrupt and, in the Jews' case, terrible constriction, as from pogroms, forced relocations, and mass murder.

The major reason for the turn in Ashkenazi fortunes was the Crusades, beginning in the late eleventh century. Seeking to retake Jerusalem for Christianity, armies from northern Europe marched against the infidel Moors and, along the way to the Holy Land, took cruel swipes at Jewish communities. Local violence against Jews was legitimized during the feverish period of the Crusades; many Jews were killed. In the twelfth and thirteenth centuries, England, France, and Germany expelled their Jewish populations altogether. Those who filtered back to Germany a century or two later were made to wear yellow badges on their coats or caps so that the authorities could keep an eye on them. Throughout, the educated Jew would serve the local prince as treasurer, banker, doctor, scholar, well aware that he could lose all when the climate changed.

Spurned, the mass of Ashkenazim gradually migrated east to Poland, where they felt welcome, and north to Lithuania, a land whose lakes froze black in winter, imagine, and Russia, such a far cry from the stony hills of Palestine. In their new territory the Jews rebounded from persecution. This would be the Ashkenazi homeland for five centuries. Shtetl life began, the thriving era of the Jewish village and Jewish urban quarter and busy synagogue. The village of Ostrog, from which Harry Ostrer's name probably derives, was one such active village. Ownership of Ostrog floated among Russia, Ukraine, Poland, and Lithuania—the geopolitical title didn't matter to Jews because they could do nothing about it. They paid their taxes and went about their business. Every now and then a Cossack raid or catastrophic fire would destroy the Jewish sector, and they would rebuild it.

Twenty thousand Ashkenazim increased to two million. In 1791, Russia, confining Jews behind the Pale of Settlement, closed the borders of the eastern Ashkenazi homeland. It resulted in overcrowding, poverty, and additional genetic concentration. When, as part of a move to Russify the Jews, the czars instituted a military draft, many young men went on the move. The rovers included one Shmuel Ostrer, who was Harry Ostrer's great-grandfather. Shmuel's unsettled descendants continued to migrate, first to Vilnius, the Jewish hub in Lithuania, and thence to Boston, Massachusetts, early in the twentieth century.

Despite its stops and starts, the world's Ashkenazi population attained ten million by 1900. The geneticists who study the group often comment on its steep growth curve—they use the term demographic miracle. They can tell that the population grew fast because harmful mutations emerged from the bottlenecks, a downside of the miracle. For example, 185delAG, occurring in one in a hundred Ashkenazim today, could not have become so prevalent if the population hadn't formerly been so small. The principal mutation for Tay-Sachs disease, which is carried by one in twenty-five Ashkenazim, arose from a handful of founders. All told, Jews of Central and Eastern European extraction are host to forty or more genetic disorders, according to research by Ostrer and other scientists. These are medical conditions that larger ethnic groups do not have, or do not share to the same extent.

Whether your name is Cohen or Wheelwright, you could have one of these disorders, Harry said tartly.

True, I said, but almost certainly the mutation behind it would be spelled differently from the Jewish form, and the probability of my inheriting or transmitting the disorder would be lower.

To be very clear, the Jewish genetic diseases are rare. They're asterisks when compared with the diabetes, hypertension, heart disease, cancer, etc., afflicting all population groups. Also, it is important to distinguish the dominant disorders, such as heritable breast and ovarian cancer, HBOC, which is transmitted by one parent, from the more numerous recessive disorders, such as Tay-Sachs disease, which entail two parental carriers and a one-in-four chance of inheritance. The genotypes outnumber the phenotypes, a short way of saying that healthy Jewish carriers of a recessive mutation greatly exceed the number of those who actually get

sick. The point to return to is that inbreeding, founder effects, population bottlenecks, and genetic drift, the instruments of inherited disease,
happen to be the very elements of Jewish struggle and survival.

The situation improved when the Jews of Western Europe were
let out of their urban ghettos in the nineteenth century. The assimilated branch of Jewry met European societies halfway, embracing and
enhancing the secular cultures. The same accommodation happened in
the United States. It was the immigration of the poorer, more insular
Ashkenazim, arriving by the millions from the East, that prompted the
so-called Jewish problem in Western societies at the turn of the twentieth century. The debate in the West focused on how many to let in and
how to absorb them. This time, Jews rankled not so much because of
their religion but because of their race.

Current thinking regards race as a hollow category, a grab bag of
traits that distract human beings from their commonality. Dismissing
biology altogether, some academics call race a social construction tacked
together from cultural traits and maintained by cultural expectations.
Race begets only racism, in this view. At most, race should be about
identity, critics say, and they brandish genetic studies: If you take two
people of different races, any two people from anywhere on the globe,
and compare the several billion bases of DNA that make up their respec-
tive genomes, you will find the two genomes 99 percent the same.

At the beginning of the twentieth century, however, most of the best
minds in science, religion and politics, Jewish minds not least, believed
that the human races were as separate as feudal castles in the medieval
countryside. It was a long-standing medieval idea. What you saw on
the outside of different peoples, be it skin color or behavior, reflected
the innate, blood-borne indicators of their race. You could, by interbreeding, cross the moats that nature had built around racial castles,
but the results were usually bad. Darker peoples were held to be less fit
than the Caucasian standard-bearers. Whether the basic human types
had been set in stone by God or had diverged or degenerated from
whites because of environmental factors, the divisions of mankind were
believed to be grave and deep.

Ashkenazi Jews in both Europe and North America endorsed the idea of their race. The issue surrounding the Jewish, or Semite, race was where it fit into the fuzzy hierarchy. Were the Semites a secondary sort of Caucasian? Or a somewhat higher strain of African, as the anti-semites in anthropology maintained? Or were they their own Middle Eastern race? Race science having become respectable, Jews too were interested in defining Jews—for it was dangerous to cede the interpretations to others.

The boiling intellectual and political currents of the turn of the century, including Zionism, social Darwinism, and eugenics, are not going to be entered here. Enough to say that nearly all scientists endorsed the biology of race; the problem was defining the human types with any sort of precision. The established procedure was to measure people—to measure them strenuously in all their external aspects, not just their racial pigment and hair texture but also their eye color, height, weight, arm span, muscle structure, lung function, and so on, and particularly the width and length of their heads, which produced a ratio called the cephalic index. You might combine these measurements with looser indicators like nostrility, the fleshy placement of the nostrils high on the face. Prominent nostrility was a giveaway of the Ashkenazi Jew. Although this trait was first documented by Jewish scientists, who wrote it off as minor and unreliable, their antagonists in the profession regularly turned the Jewish scholarship to their own ends. The trouble, obviously, with the morphometric approach was that the physical measurements weren't consistent within a single family, let alone across a racial type. Mindful of human beings specifically, Charles Darwin had warned that it was rash to attempt to define species using inconstant characteristics.

Race science had grown into a large, shambling collection of phenotypes, crying out, in the brave new world of Mendel, for the underlying genotypes. Mendel, at last Gregor Mendel. Working with pea plants in a monastery half a century earlier, Mendel had discovered and published simple rules of heredity. After a curious lag in recognition, the rules of Mendelism were being employed enthusiastically. For biologists to be able to predict the transmission of genes and to observe phenotypic traits emerging was very satisfying, yet knowledge of the seminal material, DNA, lay as far in the future as Mendel's work was in the past. Genes were abstractions, mysterious particles of inheritance, and as abstractions were flagrantly tossed about. You could argue that the racial characteristics of Jews or Africans or Mexicans were genetically controlled, but that's all it was, an argument.

For most of the twentieth century, the best statement on the question of Jewish race was ignored. In 1911, Maurice Fishberg, an anthropologist and physician at New York University, published *The Jews: A Study of Race and Environment.* Fishberg's book made the case that Jews had been transformed by gradually interbreeding with their fellow Europeans. He had crunched the physical statistics and found that Jews in any given country were more like their gentile countrymen than like Jews living elsewhere. Whatever racial uniqueness they may have had was lost, he maintained.

Fishberg had no way to quantify the admixture that had occurred. Modern genetic evidence suggests that the rate of outbreeding by Ashkenazim was as low as 0.5 percent per generation. Does that sink his argument? No, because some eighty generations have elapsed since the founding of the Ashkenazim. Tiny genetic leaks through the dikes guarding the population could have built up over centuries and dissolved some of their phenotypic differences with outsiders.

In Maurice Fishberg's optimistic vision of humanity, people were as plastic as sunflowers, responding rapidly to changes in the environment. He collaborated with another progressive anthropologist, Franz Boas, on a famous study of the environment's effects on race. The study showed that the social conditions of America had altered the head sizes of Jews and other immigrant groups within one generation of their arrival from Europe. Fishberg thought that everything that made Jews distinctive, including the way they walked and talked, reflected the landscape of their clannish communities. Their traits stemmed from long isolation and persecution, not biology. If a type existed, Fishberg believed, it was

THE WANDERING GENE | 53

a social type residing in the spirit of the Jew instead of his body. "It is not the complexion, the nose, the lips, the head which is characteristic," he declared. "It is his soul which betrays his faith."

"Judaism prospers best under the iron rule of isolation," Fishberg wrote, adding that what was good for Judaism was probably not good for Jewish people. Therefore he was eager for their assimilation. Assimilation was the answer to the Jewish problem in the West. Yet nothing Fishberg said or wrote could derail the ravenous applications of race science, which continued to their cruel climax in Josef Mengele's laboratory at Auschwitz in 1944. It may be inflammatory and reductive to finish this history with Mengele. Moreover, it is not clear what Dr. Mengele was up to, although hunting for the internal pathways of Jewish genes was part of his assignment. If his experiments could identify the genetic components of Jewishness, the information could be used to extract members of the race from society, even when they had admixed with other Europeans. Mengele was the last researcher, if that's the right word for him, to try to classify a people by pulling apart their bodies.

Harry Ostrer took pride in his connection to Maurice Fishberg, a prophet in the scientific wilderness. Both men held posts at the New York University School of Medicine. Harry thought of himself as Fishberg's heir, and sometimes he titled an article or lecture accordingly, such as in "Maurice Fishberg's Legacy: Population Genetics in the 21st Century," a talk he gave to his NYU colleagues.

Harry was a voice in the wilderness too, but on the other side from Fishberg. He studied the genetic variants characteristic of Jews, the hardwired facts that might set them apart from other people. Race was passé, no need for him to fight that battle, but he wasn't afraid to address the overlap of ethnicity and biology, politically incorrect though that might be.

He launched an ambitious study of Jews' origins and migrations called the Jewish HapMap Project. When he had time, he went into the field and collected their DNA himself. I am in Thessaloniki [Greece] recruiting for the Jewish HapMap Project, he e-mailed. As you know, it was a majority Jewish city prior to WWII. The small Jewish community now is comprised of Holocaust survivors, their children, and their grandchildren. Sometimes the tears appear when I draw blood from an arm on which a number has been tattooed and read a family history questionnaire that ends suddenly at Death Camps.

In 2010, Ostrer and his colleagues published their analysis of seven Jewish populations from Europe and the Middle East. They detected genetic threads tying all of the populations together, regardless of the distance separating the groups today. Jews were distinguishable from their longtime neighbors of other creeds by means that Fishberg would never have thought possible. The work made an international splash. When a critic in Israel complained that Hitler would have been very pleased, Harry replied bluntly, We can tell who the Jews are genetically. Privately he wondered if he had produced the modern version of Nazi race science. He could be quick to second-guess himself.

For what they're worth, these further attributes of Harry Ostrer are offered: mercurial, hard to pin down, a big-picture man, a synthesizer of results, prickly about his reputation, uneasy in his alliances, happiest when working hardest, a thousand balls in the air, proud to be a Jew, proud to be an American, a *macher*, a tough kidder, a softie who hated seeing others in pain (he winced over and over when I told him what Shonnie went through), a terse, honest, invaluable guide.

From the beginning of genetic science, some of its specialists (not Ostrer) have wanted to uncover a higher purpose in the function of genes. The mindless honing of DNA on the assembly line of evolution, the arbitrary forces of genetic drift, and especially the creation and persistence of disease mutations were disquieting to many in the field. They wondered whether the genetic instructions to the body had a purpose even when the body suffered. While most medical geneticists focused on diseases and deficits, the theoreticians of carrier advantage sought to flip the work in a positive direction.

The familiar example of carrier advantage is sickle-cell anemia, a blood disease seen mainly in Africa. It's a recessive disorder, meaning that it takes two copies of the sickle-cell mutation to bring about the disease in carriers. However, bearers of *one* copy of the abnormal gene gain some protection from malarial infections while being spared the worst effects of the blood disorder. Evolution, it appears, has favored the continuation of the sickle-cell trait in human beings who have had to contend with mosquitoes and malaria. The math of the trade-off is straightforward. For every person who perished from inheriting a recessive illness, two or more healthy carriers went ahead with their fitness and survival enhanced.

Tay-Sachs disease has served as the analogous disorder for Ashkenazi Jews, that is, for the scientists attempting to divine its recessive utility. By all appearances, Tay-Sachs—named for two doctors who described it at the end of the nineteenth century—has nothing in its favor. It was and is a very grim neurological condition. Affected infants start life healthy but begin to lose muscular control at six months. Their heads loll helplessly and their vision dims; after a period of convulsions, they slide into coma and death. How might the healthy carrier have an edge over those lacking the gene?

In Maurice Fishberg's day, Tay-Sachs was known as amaurotic family idiocy (a name that wouldn't fly now for any disease). Fishberg thought the disorder might be inherited, but he would not attribute it to inbreeding, which he considered an ill-founded slur by prejudiced gentile doctors. Non-Jewish physicians were in agreement that the Jews of the urban ghettos were sickly, and prone to nervous disorders, and generally of inferior constitution—except, curiously, when they were exposed to tuberculosis and cholera. Just as Negroes seemed to resist malaria better than other races, Jews seemed to carry a protection against tuberculosis. This century-old observation was short of data and has been relegated to the urban-legend category, but back then it made for lively scientific discussions. Adopting a Darwinian line, Fishberg averred that constant exposure to tuberculosis in crowded European cities had weeded out the less-fit Jews—those excessively predisposed to infection, in his words—and that the protective result had carried over to the milieu of New York City.

Fishberg knew nothing of DNA and carrier advantage. But half a century later, evolution-versed geneticists understood the transmission of recessive disorders and the potential of the healthy carrier. That's when the possibility that Tay-Sachs might protect against TB was brought back and investigated. The inverse association between the two conditions was anecdotal: In surveys of Ashkenazi patients, accounts of tuberculosis in the patients' families didn't overlap with histories of Tay-Sachs. People remembered that their grandparents in Eastern Europe had suffered from one disease or the other but rarely both. Geographically, too, the disease histories didn't overlap, as if one condition might be excluding the other. As an additional factor, Jews seemed underrepresented in sanatoriums for TB patients. Well, it was a good story as far as it went, but no biological mechanism was proposed, and the Tay-Sachs–TB association lost traction over the years.

A new theory was advanced in 2005, this one having to do with Jewish IQ. Anthropologists in Utah published it in an obscure journal, and the *New York Times* jumped on it. Since Jews had always relied on their intelligence, Tay-Sachs mutations might have spurred brain growth, the researchers wrote. The Ashkenazim who carried a Tay-Sachs mutation, or a mutation for three somewhat similar disorders, might have been favored with greater mental development than other people. In the pressure-cooker of the ghetto, the smartest men would have done best in business, and would have spawned the largest families, which would have spread their intelligence genes. Hence the many Jewish Nobel Prize winners in the twentieth century. The rare recessive diseases were kind of an accidental by-product of Jews' mental success.

The Utah research team also managed to work 185delAG and two other Jewish BRCA mutations into their theory about IQ. Previously, scientists had speculated that during periods of famine the female carriers of a BRCA mutation may have gained an advantage in fertility or lactation, which could have helped their offspring to survive. The Utah anthropologists noted that BRCA1 and 2 were active in neuronal cells, where they appeared to take part in restraining cell proliferation. A neuronal cell normally has two working copies of the BRCA genes to perform this task. What if having a single defective copy slightly unleashed brain growth so as to be advantageous to Jews? Again, all well and good. You couldn't prove it *didn't* happen that way unless you conducted an experiment, and nobody was proposing to count the brain cells or measure the IQs of BRCA carriers, men and women who already were quite preoccupied with the gene's drawbacks.

The Utah researchers—Henry Harpending, Gregory Cochran, and Jason Hardy—were audacious to put an evolutionary spin on Jewish braininess, and they took a lot of criticism for doing so. Overshadowed in the IQ controversy was the implicit desire to discern order and progress in DNA. The forces of evolution, though blind, were not heartless, according to this view. Natural selection was a much more appealing explanation for the deleterious mutations than inbreeding, founder effects, bottlenecks, and drift, the Four Horsemen of genetic variation.

Dismissing the Utah paper as poor, Harry sought to place it in a broader context. Historical phenotypes for Jews have been bandied about, both adaptive like this one and maladaptive, he said. He and other Jewish scientists labored under a double-edged sword. With the Holocaust still echoing, Jews didn't want to be viewed as inferior but it wasn't good to be seen as superior either. As he talked, he switched casually between the genetic and cultural sides of the story, now presenting one, now trying on another, a matador wielding a prayer shawl instead of a cape.

So the Ashkenazi Jews endured their severest bottleneck, the Holocaust, and after World War II started to grow vigorously once more. Race science was dead; the science of genetics went forward. The genocentric age that is upon us will be discussed in a later chapter; a few more things must be asked about the Holocaust first. Was Jewish DNA rearranged because of it? Geneticists don't know. It was not like earlier bottlenecks because the population was so much larger in the twentieth century, and it had spawned separate subpopulations. Two of these, the Romanian Jews and the Lithuanian Jews, the Holocaust removed from the map. Most probably, the DNA has been constant and only the raw numbers shifted—the millions who were terribly lost.

A large part of the world's 185delAG carriers perished as well. Regarding the remainder, not all of them Ashkenazi, Shonnie Medina and members of her family being highest in mind, consider the likelihood that every one is a descendant of a Judean who walked in the desert twenty-five hundred years ago. There's a technical term for them: identical by descent. In a double sense these individuals were born to suffer: first from disease, if the full weight of the gene prevailed, and second from the biblical onus of the dispersed and chosen people.

Identifying the 185delAG mutation could be more difficult in the future because of intermarriage and assimilation. The American Jewish community is the largest, most powerful Jewish community in all of history, including the era of the Second Temple, but half of young Jews are marrying outside of the faith, and the religious upbringing of their children is uncertain. Worldwide, the number of Jews has barely increased since 1970, in the face of growing numbers of people of other faiths. Maurice Fishberg observed that Judaism weakened when Jews were fully tolerated, just as persecution always made them strong. In the twenty-first century, it is the culture that is under threat of dissolving, or perhaps it will retreat to the strongholds of the Orthodox Jews and to Israel, while the mutation travels like a wind-blown wave, not weakening or slackening. For BRCA1.185delAG is a dominant force acting by itself, unlike the recessive agents, whose complete effect requires two partners.

The Jews' place in Western society remained a mild dilemma for Harry. The research by him and others showing a mix of European and Middle Eastern DNA hadn't made the dilemma go away, only embellished it. We're just like the rest, was the way he put it, versus We are special.

We sometimes say we're in a golden age, he mused. Golden for its tolerance of Jews and for the scientific studies of the Jewish population. But things could change. Golden ages end. We could go through tough times again.

Chapter 4

EL CONVENTO

San Luis, the hub of Culebra, claims to be the oldest town in Colorado, finessing the fact that San Luis was part of New Mexico Territory when the town was established, in 1851. The town and its streets feel very small because the natural features and religious aura of the place are so large. A century-old church, the biggest building in town, anchors the central plaza, and a modern chapel, gleaming and marmoreal, overlooks San Luis from the steep mesa hugging the town. You are never out of sight of one structure or the other, the effect of which is to make the place bristle with symbols of Roman Catholicism. The vaulting sky and mountains dilute the impression somewhat.

But now hike up the half-mile-long path braiding the side of the San Pedro Mesa. The entire path is a Catholic shrine; it is the main tourist draw of San Luis. Completed in 1990 and topped by the white Chapel of All Saints, the Stations of the Cross Shrine produces its impact slowly. Every forty yards or so, you pass a bronze sculpture of figures closely engaged in the incidents of the Crucifixion. The sculptor, Huberto Maestas, is said to have put Culebran faces on the three-quarter-scale figures. Burnished on the drab hillside, the stations could plausibly mark the way to Golgotha. Loose rock, prickly pear cactus, sagebrush, and juni-per adorn the path where Christ trudged, stumbled, and fell. Switch-backs force you to double back and come face-to-face with yourself, and by the time you have reached Station 11 you may be short of breath and quite unnerved. Although the Gospels don't elaborate on the nailing, of course it makes sense that Jesus would have been laid on the ground, the rough wood beneath him, in order for the man who did the job to swing his arm high and pound the nail through Jesus' wrist.

So he hangs above Culebra, as the morning light descends on the mesa. The viridescent Valley lies all around, its vestige of snow wicked away by the cold April sun. In the foreground of the western mountains, the lower, darker mesas will be illuminated when the sun, clearing the Sangre de Cristos, discovers them one by one on the plain. Willa Cather wrote about the lit terrain, ". . . those red hills never became vermilion, but a more and more intense rose-carnelian; not the colour of living blood . . . but the colour of the dried blood of saints and martyrs."

Happily, the sculptor has made a fifteenth station, which is set apart from the other scenes on the mesa. Walk around to the north side of the handsomely domed chapel. The final bronze depicts Christ ascending from the top of his cross like a diver buoyantly regaining the surface. A whoosh of air is about to fill his lungs, the bends of death be damned.

You are sitting above the circle of the earth. About one hundred miles long and fifty to sixty miles wide, the San Luis Valley is reputedly the largest alpine valley in the world. Imagine an area the size of Connecti-cut lifted into a deep blue sky and surrounded with snow-capped moun-tains. Because the desert air foreshortens the distances, any sight line quickly runs into an unscalable wall. The Valley is a time capsule, cut off from the concerns of the rest of the world. A freeze-dried religiosity from sixteenth-century Spain persists in the landscape, not degrading but also not penetrating all the way through to church attendance as it used to do. The Valley is a mountain henge, a sealed system. Growing up in this enclosure like one of Mendel's pea plants, Shonnie Medina blossomed according to the stern rules of her Spanish lineage, although the rules were bent by a sharp turn in the family culture, and then by a single gene.

When you have come down from the mesa, and if you are still game, take yourself across the plaza to the San Luis Museum. Go straight upstairs, bypassing the first floor and the cramped, conventional exhibits of Culebra's history. Few visit the tiny museum, still fewer mount to the second floor. The wooden treads groan; the light falls away and you come into a tenebrous quiet. Before you is a replica of the main room of a *morada*, where the *penitentes*, the secret brotherhood, used to meet. *Morada* means home or dwelling. The word deliberately understates the goings-on inside.

Three or four wooden benches—rough pews—face the north wall of the room. On the walls are *retablos*, simple, flat paintings that were modeled on the altarpieces of Spanish churches, and in front of the benches are primitive, life-size sculptures and mannequins called *bultos*. The *bultos* portray two dimensions of the human form plus a hint of the third, the plumpness of life painfully pressed out. There stands the bloodied Jesus, his eyes hooded, and also a hazily dignified figure known as the Virgin in Black. Another *bulto* depicts Our Lady of Death on the hunt. A skeleton with a plaster face, Lady Death stands stiffly in a square cart with an arrow pulled back in her bow. During Good Friday processions, one of the *penitentes* would haul this cart, sometimes made weightier by sand or rocks, along the road past the goose-bumped onlookers. "She is like Cupid," writes one authority, "except that her arrows bring death instead of love." Doña Sebastiana—the polite sobriquet you may use for her—scans coldly over your face in the crowd. "Sometimes as the cart runs over a rough spot, the arrow may leave the bow, and unfortunate is the person whom it strikes for it is known that he will die shortly."

A Plexiglas case contains various smaller devices of the penitent brotherhood: crosses, statues, candleholders made from tin cans, *matracas* (noisemakers), *divisas* (ribbons), and a *disciplina* (whip). The *matraca* spins in a housing like a gear with wooden teeth. When you rapidly rotate your wrist, the *matraca* makes a penetrating sound, intensifying the ritual. The *disciplina* was used for flagellation.

The exhibit's artifacts, collectively called *santos*, were retrieved from defunct *moradas* in New Mexico and Colorado. The real *moradas* were austere places that made do with many fewer icons and accoutrements than are presented here. The theatrical struggle between good and evil, sin and grace, could be enacted on the bare platforms of men's bodies. When Shonnie and Iona were little girls, they started to cry during a ceremony in the San Francisco *morada*, even though the rites by then were much toned-down. A devil of some sort jumped out of the dark, spinning his *matraca*. Marianne swore never to take her girls there again. This was not long before the family became Jehovah's Witnesses.

Downstairs into the light again, taking a couple of deep breaths. What used to be a convent-school sits a little farther along the plaza: El Convento. To enter, you pass through a courtyard-style stucco wall. The brief, unattached wall is an architectural gesture, which cannot actually have defended the modesty of the virgins inside, but the building itself could have. The two-story adobe *convento* squats on its site like a fortress. It has massive, double-hung windows and a blocky mansard roof with thick shingles and a belfry. The first occupants were the Sisters of Mercy, who lived upstairs. Downstairs, the nuns taught parochial students drawn from Culebra; for a long time Mercy Academy served as the only high school in the county. Day and night, El Convento is almost exhilaratingly gloomy. Today the Sangre de Cristo Parish operates it as a bed-and-breakfast.

Immediately on entering, the tone is set by a tall painting of Saint Cajetan. A sixteenth-century Italian reformer, Cajetan cofounded the Theatine religious order. Seeking poverty and purity, the Theatine monks were part of the movement to tighten moral standards and inject new zeal into the Catholic Church in response to Martin Luther's shocking and heretical challenge. From Italy the Theatines spread to Spain and to missions in the New World and the Far East. Though the order, in the global Catholic dominion, has waned over the centuries and is a minor one at present, the Theatines still manage the spiritual care of Hispanic Catholics in a few dioceses of Colorado. The Sangre de Cristo Parish is run by Theatines, hence the honoring of Cajetan. The priests and brothers live in the rectory across the street.

Cajetan, with black beard and habit, posing before a dun-colored sky, rolls up his eyes to a hovering dove, ignoring the guest who has put down his bags in the hall. No help is forthcoming, no staff except across the street. Other artworks and churchy photographs, which emerge gradually from the dark wood paneling, may try the nonbeliever. El Greco and Francisco Ribalta would be right at home in El Convento, while Velázquez and Goya might prefer to look for something a little brighter. To be fair, once you settle in, there's nothing grisly or scary about your high-ceilinged room; some barbed wire around one of the *santos*, that's about as far as it goes.

Say your prayers and go to sleep. Meditating on Christ's wounds, Spain's artists in the late sixteenth and early seventeenth centuries depicted his Passion with all the naturalism they could muster. The Protestants of the Reformation countries condemned such images as idolatry. But the mystical Catholics of the Counter-Reformation were striving to experience his suffering body to body, through which they might be joined to him soul to soul. Is that their ghosts you hear coming home from the *morada*? Or just innocent creaks in the hall?

Harry Ostrer came to stay once at El Convento. It sounds unlikely but it's true. When he came in and set down his suitcase, he surveyed Saint Cajetan cautiously. He had been forewarned about the intensely Catholic atmosphere of the place.

You know, he said after settling in, we all used to be Sephardim.

The Sephardim are the Jews of the Mediterranean basin. It will be recalled that Jews dispersed about the Mediterranean following their expulsion from Jerusalem in the first century CE. Somewhere in the neighborhood of Italy, an offshoot occurred. The Jews who begat the Ashkenazi population and Harry's family went north to France and Germany. The Hebrew traders and travelers who continued west to the Iberian Peninsula, going by way of North Africa, became the founders of the Sephardic branch of Jewry. Sepharad was the name in the Hebrew Bible for a place of uncertain location. To the Jewish migrants, Sepharad seemed to be just that place. The Romans called it Hispania (Spain).

During the Middle Ages, the Sephardim were the world's largest and most prosperous tribe of Jews. From this population came Maimonides, Baruch Spinoza, Moses Montefiore, Emma Lazarus, Pierre Mendès-France, Benjamin Disraeli, and Jacques Derrida, not all of them practicing Jews. The Sephardim, though today they are outnumbered and outshined by the Ashkenazim, object to the too-easy conflation of Ashkenazi and Jewish, as the previous chapter has done. Mizrahi Jews, who traditionally lived farther east, might rightly voice the same complaint. But this is about long-ago Sephardim and their lost branch in Spain: the late-medieval, freshly minted, sorely missed New Christian branch of Jews, which included Tomás de Torquemada, organizer of the Spanish Inquisition; Saint Teresa of Avila, the hard-headed Catholic mystic; possibly Miguel de Cervantes, author of *Don Quixote*. And many more—down to the family of Shonnie Medina.

To resume the story of Sepharad: Ironically, the best thing that happened to the Jews while they were living in Sepharad was that the Muslims from North Africa seized Spain during the eighth century. Prior to the Arab conquest, Christians in Spain had forcibly baptized Jews, among other punishments; under the Moors' rule the persecutions and forced conversions ceased. Jews and Christians not only were tolerated but also were allowed to thrive as long as they paid taxes and refrained from proselytizing. Cosmopolitan Sephardic Jews spoke Arabic, wrote poetry, and contributed their learning to Moorish libraries. Beside the Great Mosque of Córdoba they strolled among the orange trees and felt safe. Jewish people today look back on this as a golden time, rivaling the era of the First Temple.

Golden ages end, insofar as the Jews are concerned, not with a whimper but with a crusade. As the Muslim emirs, starting in the eleventh century, gave up ground on the Iberian Peninsula to Catholic princes and their armies, hostility to Jews increased. Initially the Sephardim suffered collateral damage in the crossfire between the two warring powers. Subsequently, in the territories where Catholicism was reestablished, Spanish priests, backed by riotous mobs, demanded that Jews and Moors convert. Tens of thousands complied, under pain of death.

After 1450, the Spanish kingdoms became less concerned with forcibly converting Jews than with ensuring that the prior conversions had been sincere. This was the motive for establishing the Spanish Inquisition. The Inquisition's principal targets were crypto-Jews—Jews who claimed to have converted but continued to practice their beliefs in private. No doubt the crypto-Jews were just a minority of the *conversos*, a small minority, but they were tied to the genuine converts through family and business, and they communicated with the Jews in the ghettos who had not converted. Such Spaniards led double lives. Secretly they lit candles and cleansed themselves and their clothing on Friday nights before Shabbat, then they showed up for Mass on Sunday like everyone else. For eating pork, when by rights they shouldn't, they became known as *Marranos* (swine). The Inquisition sought to root them out, this fifth column, by torture and the stake if necessary, since it was a capital offense for a Christian to backslide and be a Judaizer. If convicted, the crypto-Jews were given a chance to kiss the Cross, after which they would be garroted. If they refused the crucifix, they'd die more slowly by burning.

Race and blood fueled the campaign. Fifteenth-century Spain instituted *limpieza de sangre* (purity of blood) statutes, not unlike the Nuremberg Laws of Nazi Germany. The *conversos* and their descendants were deemed New Christians; Old Christians were the Spaniards whose racial heritage was pure. The Old Christian looked through the shellac of conversion and perceived a racial stain. Barred from municipal positions and guild memberships, some of the New Christians fudged their genealogies in order to demonstrate that their blood was free of foreign taint. Others bought certificates or bribed their way to Spanish purity, and still others married into it, until the Church blocked marriage between *conversos* and Old Christians. The Inquisition therefore became expert at examining the pedigrees of prominent citizens.

In 1492 the Catholic monarchs, Ferdinand and Isabella, having united Spain, drove the last of the Moors from the Iberian Peninsula. The country's wobbly identity needed to be consolidated, so the Crown lowered the boom on the remaining Jews. According to decree, all Jews who had not or would not convert to Christianity were to be expelled from Spain. The royal Inquisition enforced the order. As many as 100,000 Sephardim converted, and as many as 150,000 departed, first for Portugal, which offered them sanctuary for a few years, and then for more distant parts of the world. The Jewish people once again were dispersed and weeping. The Moors faced the same choice shortly afterward.

Also in 1492, Christopher Columbus set sail, which has led to speculation that he might have been a crypto-Jew or had reason to fear the Inquisition. Columbus is known to have had New Christians on his voyage. But then ethnic mixing in Spain had been going on for so long that in any historical collection of Spaniards, as on ships to the New World, you can uncover the genealogical tracks of Jews.

Historians have teased out code words for Jewishness from ostensibly Christian texts, Cervantes's *Don Quixote* being primary material. For starters, La Mancha, the name of the knight's home province, means stain or spot, possibly a playful allusion to the hero's (or the author's) race. Additionally, it is pointed out, Cervantes's father was a doctor (medieval physicians often were Jewish); his mother may possibly have been from *converso* stock; Don Quixote's library is publicly burned (i.e., to expunge the People of the Book?); the accounts in the novel of priests, penitents, and the Catholic Church are unfailingly ironical, if not snide. Finally, why is it that whenever Sancho Panza refers to himself as an Old Christian—he seems to pat himself on the back for his religious lineage—why does the knight never say anything in return?

The double consciousness and zigzagging identities of the *conversos* are the focus of the postmodern historian, rather than their anguished need to worship. For the Spanish Inquisition, the persistence of Judaizing, decades after a family's conversion, confirmed that there was a biological imperative to being Jewish. In one sense the medieval authorities were right about the special nature of *converso* blood. In a 2008 study titled "The Genetic Legacy of Religious Diversity and Intolerance: Paternal Lineages of Christians, Jews, and Muslims in the Iberian Peninsula," a group of Spanish, Israeli, and British researchers looked for tracks of the *conversos* by examining the Y chromosomes of more than a thousand male inhabitants of Spain and Portugal. The Y chromosome of males is convenient for ancestry studies because, handed down from father to son, it barely changes over the centuries. The Y chromosome provides a narrow, very distant view of a man's past. According to the study, contemporary Iberians evince, as a group, 20 percent Jewish ancestry.

Put another way, if you assembled a crowd in Toledo, a city in La Mancha, and pulled out the DNA of a typical Toledan, there's a one-in-five probability that he would carry Sephardic ancestral markers. Understand that there's no single definitive marker on the Y chromosome that proves Sephardic ancestry. The calculation of 20 percent was based on genetic frequencies drawn from collections of people. Previously, scientists had taken Jews of proven Sephardic stock, most of whom were living outside of Spain, and from these descendants of exiles came a DNA standard, a baseline Sephardic formula, that was used in the new study. The Sephardic markers found in the test subjects were compared with the DNA of the parental stock of Iberia, which was assumed to be Basque. By the way, besides their quotient of Sephardic blood, Spanish and Portuguese men are 10 percent Moorish, according to the study.

Although you may not uncover the breast-cancer mutation 185delAG in your random sample of Toledo, 185delAG, as noted, is nearly surefire proof of Jewish ancestry (a broader, older category than Sephardic). Sepharad was the mutation's westernmost stopping place in Europe after its carriers migrated from Judea. The 185delAG marker does illuminate the individual, one part of him or her, unlike the probabilistic markers used in the Iberian ancestry study. Less insular than the Ashkenazi tribe of Jews, the Sephardim do not appear to have built up the same prevalence of the wandering mutation. The frequency is impossible to know for sure, since genetic and medical research is limited for this group in comparison with the Ashkenazim. Not numerous anywhere since their great exodus in 1492, Sephardim are widely scattered about the globe. However, one fact about the gene in Spain is revealing. In a 2003 survey of breast-cancer mutations affecting four hundred high-risk families, 185delAG topped the list of the BRCA1 mutations. This finding meshes with the Sephardic ancestry work.

Dogs bark at night in the yards around El Convento. The dogs start around 10 p.m., knowing one another well. It is a ruminative, conversational barking that lasts until midnight, the sound seeping in from the tall, heavy windows. One animal seems to take the melody, the others the contrapuntal parts. Declaiming, dreaming about the beautiful shepherd and the rabid dog. In the dualistic landscape of Culebra, a human being must pass between good and evil, light and dark, treachery and faith. Who knew, when the dog was let into the corral, that it had gone mad?

Spanish Catholicism in the sixteenth century looks unbelievably chauvinistic and harsh. As if making a last-gasp stand against the Renaissance and the Scientific Revolution, Spain was the arch-reactionary of Europe. Reactionary in rejecting Protestantism, in battling Islam, in rooting out Jews, in conquering and converting Native Americans, and yet each of the cleansing, cauterizing campaigns was conducted in the name of Jesus Christ, and that is the hardest thing about those Catholics to understand, let alone identify with. Present-day sensibility is dismayed by the medieval Catholic mind.

Witchcraft and enchantment have to be factored in. Awareness and respect for the dark side of their minds are key to understanding how these Catholics felt, or at least to feeling how they must have thought. Prescientific Spaniards believed in witchery and in the witches' evil taskmaster, Satan, who was walking about half-seen on the earth, fleeting in the air, and interfering in human affairs. The devil sent witches to ruin people's crops, sicken their children, and torment their bodies and minds. Outside the arena of popes, princes, and prelates, the theological politicians who directed the Inquisition and other historical forces, the lion's share of Catholics, ignorant and afraid, channeled their anxieties through folk religion. Folk Catholicism was a superstitious, populist reaction to the war that was underway between Christ's forces and those of the devil.

Superstition—arguably the wrong word. Is belief in Christ and the holy works of saints superstition? As science gained strength in Europe, it exorcised the evil half of the dueling pair, which left God solely in charge of the supernatural realm. In the meantime, since the boundary between the natural and the supernatural did not exist or could not be located, armoring yourself with Christian superstition was a normal, perhaps even rational, response to the combat between God and Satan. Similarly the crucifixes, rattles, relics, and effigies on display in the San Luis Museum's *morada* were church-sanctioned shields against the dark side.

What would the Inquisition have made of Shonnie and her coreligionists in the Jehovah's Witnesses? Their rejection of political authority would have been a problem, though they would not have been subversives. Their Christian superstitions (they reject Darwin and evolution) may well have passed muster. The Witnesses do fundamentally accept the reality of Satan, without believing in his witches; they think the devil has adjusted his methods to the scientific age. It is time to debrief a Witness.

Guests at El Convento take breakfast in a small, handsome dining room. Coming up the walk to have breakfast this morning is Chavela (Chavelita) Medina Salcido, who is Joseph's sister and Shonnie's aunt.

Cool and composed, Chavela was the first in the family to take up the Jehovah's Witnesses, about thirty years ago, and her example encouraged the others. She was the second in the family, after Shonnie, to find out that she'd inherited the breast-cancer mutation. Her expression as she got out of her car today was pensive. She volunteered that she didn't like to come to San Luis because the memories made her sad. My memories of childhood, she said unexpectedly. My father was an angry . . . abusive . . . alcoholic who beat my mother every day.

Chavela's sister Lupita has a similar memory of her father. Why was he so mean? Lupita wondered. He beat her up for no reason, for jealousy, I guess, Lupita said. But Joe U. Medina had a generous side as well, according to Joseph, the eldest and most loyal son, and Wanda, another sibling, agreed. He was kindhearted, he'd give you the shirt off his back, she said. He was just—who's that guy?—Jekyll and Hyde when he was drinking. Eventually his wife, Dorothy, fled and divorced him.

Joe U. was never a Witness. He belonged to the San Francisco *morada*, but was a listless Catholic. He seems to have cared mainly about the camaraderie of the *penitentes* and his social standing in the brotherhood. That time when Marianne quit the *morada*, Shonnie clutching her right hand and Iona her left, Joe U. had been angry. He criticized Joseph, saying, Smack her and make her come back. My father-in-law, Mr. Macho, sniffed Marianne at the memory. As a boy, Joseph had been beaten by his father, and as a teenager he had paid others in kind, but he wouldn't bad-mouth his father today and he never struck Marianne.

So here was Chavela Medina Salcido, cool, composed, and neatly dressed for the appointment. A waitress in Alamosa, aged fifty-one. Five children from two marriages. A grandmother through her eldest child, Shannon. Became a Jehovah's Witness because she liked reading the Bible and finding out for herself what was true, as opposed to being told by a priest. That level Medina gaze, which soaks you up without being impolite.

About her niece, Chavela said, She was gorgeous, beautiful, tall. Shonnie was the perfect model of a young lady. When she'd be going door-to-door, she'd be well dressed. And bubbly. She loved talking to people about the Bible.

Well, how do you handle going door-to-door when people are rude to you?

You shake the dust off your feet, shrugged Chavela, quoting Luke 9:5. The part where Jesus calls the disciples together and, as Luke puts it, gives them authority over all the demons and the power to cure sicknesses. As a tough-love initiation of the disciples, Jesus sends them out in pairs preaching from village to village, and they have to go with absolutely no possessions or money. They are told to depend on charity for their food and shelter. They shake dust from their feet to be a witness, or in other Bibles, a testimony, against the people rejecting them. The dust on their feet was the only personal item they could spare. Franciscan friars and Carmelite nuns as well as Jehovah's Witnesses have zeroed in on this part of the Gospels for instruction on evangelism.

Chavela shook her head sadly when she was asked about Shonnie's cancer. You wonder why something like that happens to good people, she said.

Do you believe that a disease like that was God's will?

She sat up straight in the dark dining room. All the illnesses that we have, she replied, they don't come from God. We know that Satan targets good people sometimes. To discourage them and their parents, so they'll blame God. It's not God. God doesn't try anybody with evil.

What about Job?

Chavela spoke without hesitation: God gave Satan permission to put Job through his trials. God told Satan, Just don't take his life.

The Book of Job is the only one in the Hebrew Bible to probe the problem of deliberate evil. Well versed on Job, Jehovah's Witnesses point out that the case was exceptional—God decided to withhold his protection from this impeccable man because of Satan's very serious challenge. Sadistically Job was made sick but through fidelity to Jehovah he recovered. He prevailed and was rewarded. As far as the rest of humanity is concerned, Witnesses believe that the origin of sickness, like the origin of sin, can be charged to Adam and Eve. Those two humans had perfectly healthy, no, immortally healthy bodies until they disobeyed God. Then they became defective, so a Witness document states, and got sick and died, and passed on the imperfection to their children. Death and sinfulness are hereditary. *What Does the Bible* Really *Teach?*, a booklet that Shonnie carried with her door-to-door, likens mankind's original sin to "a terrible inherited disease from which no one can escape."

In 2002, Chavela was prodded by Marianne Medina, her sister-in-law, to meet with Jeffrey Shaw, the genetic counselor. The two women drove to Colorado Springs together, each bringing their daughters. Marianne didn't merit BRCA testing because the breast and ovarian cancers were on Joseph's side of the family. Her daughter, Iona, manifestly was at risk because of Shonnie, but Iona asked for more time to decide. Shannon, Chavela's daughter, was advised to wait for her mother's test result, because Shannon would be a candidate for testing only if her mother was a carrier.

Of the four, then, only Chavela went forward, and after providing a blood sample she learned that she was positive for 185delAG. She had thought it best not to use her real name on the test application, in case there might be trouble with her life-insurance policy in the future. A year later, when she had a hysterectomy due to an unrelated gynecological issue, the surgeon recommended that Chavela have her ovaries removed at the same time, since BRCA carriers have a much greater chance of developing ovarian cancer than other women do. Removing the ovaries lessens the chance of breast cancer too. Chavela agreed and immediately was plunged into menopause, the normal complication after the operation. Nobody told me about that, she said.

I have the gene, Chavela continued calmly. But I haven't had [breast] cancer. I don't know what I am going to do. Because I've had many children—here she named her daughter and her five sons in order—I was told I may not get it. My mom had fourteen kids and never had cancer, and maybe that's why. Shaw, the counselor, evidently had told Chavela that pregnancy and breastfeeding disrupt a woman's estrogen cycles in a protective manner. The breast tissue receives fewer doses of estrogen overall, and that's good for warding off tumors. Conversely, it might be mentioned, this being El Convento, that nuns, not bearing any children, have higher rates of breast cancer than other women. The observation about nuns and breast cancer dates to the time of Saint Teresa of Avila.

The big counseling session by Shaw with the Medina-Martinez clan had taken place at T-ana's the day before. The meeting yesterday made me think I should be prepared, Chavela said. A lot of times we don't think it's going to happen and we push it away. I have been putting off getting my mammogram. Maybe we should have a girls' day out and all get screened. Shonnie's aunt was saying the right things, and if a little slow to act on them, she had proceeded the way that scientific medicine tells women to do when they are surrounded by heritable cancer. Talk to your relatives, consult a genetic counselor, get tested, have surgery to reduce your risk, stay vigilant; but then Chavela threw a curveball into the proceedings.

With the same rational gaze, she explained that she had sought extra protection for her health with a machine called the Bio-Enhancement Feedback Unit. The BEFEU [*Beef-ee-you*] machine, she called it. I have a toxin machine, she said. You put your foot in a bucket with water and black, oily stuff comes out of your body. Internet research revealed that the devices also go by the names Aqua Detox and Aqua-Chi machines, or ionic foot baths, and by the time you pay for the copper rings and add-ons and extra filters, you have run up some real money. As Chavela indicated, you plug the thing in and sit there until the water around your feet discolors.

It was discouraging to hear about the BEFEU because it evoked Shonnie. During her sickness, Shonnie hadn't used that particular machine but she wouldn't have been skeptical of it. Recently her aunts Wanda and Lupita had borrowed the device. The three Medina sisters were passing the BEFEU among their households, two of them doing it for insurance, but Wanda, who might well have a tumor, was using it for treatment.

Sensing a chill across the table, Chavela knew a question was coming about Shonnie and her unconventional cancer therapy. Either way it wouldn't have worked, she said flatly. It wasn't the alternative medicine. It was not catching it in time. Chavela believed that the tumor had simply spread too fast.

Around midnight, after the dogs have concluded their conversation, the cattle can be heard lowing from the *Vega*, the communal pastureland of San Luis that is watered by the *acequias*. The cows sound uneasy.

Saint Teresa of Avila wrote that learning to pray properly is like watering a garden with lighter and lighter effort. At first it's heavy, as you must lift and haul the water. But when you attain what Teresa called the third stage of effort, the spiritual schlep is over and prayer flows as easily as from a stream or spring, "although some [work] is required to channel the water the right way. But now the Lord wants to help the gardener so badly that he almost becomes the gardener." The fourth and final stage is soaking rain, just drinking God in to your parched world, if that joy can be imagined.

And still later, in a nameless hour, a chorus of coyotes, yipping and drifting closer to the *Vega* from the mountains. Joseph says that the coyotes can be triggered by the siren of a police cruiser, chasing miscreants into the Culebra night.

Teresa de Cepeda y Ahumada, a happy girl, a good Catholic girl, grew up in the Spanish city of Avila during the first half of the sixteenth century. Her paternal grandfather, who had converted from Judaism, had done well in business in Toledo. Later he admitted to Judaizing, coming forward voluntarily, and for punishment the Inquisition paraded him through the streets with his children, all dressed in shameful yellow. In the next generation, Teresa's seemingly shell-shocked father married an Old Christian and kept a low profile in life. Teresa herself grew up with verve, confidence, and wealth, the stain of her ancestry having been wiped away.

Teresa was very good-looking, and coquettish, and she loved to show off her form at dancing. Her feet especially were admired, even after she had become a nun. For all the while that she had sparkled socially, she felt a religious calling, and when it came time for her to be married off, headstrong Teresa defied her father's wishes and entered a convent in Avila. From the age of twenty until her death at sixty-seven, she spent most of her time strategically behind a wall.

Strategic, because monasticism offered a haven from the concerns and unruly demands of the day. As the Spanish philosopher Miguel de Unamuno writes, "It was, in fact, the yearning for liberty, for inward liberty, which, in the troubled days of the Inquisition, led many choice spirits to the cloister. They imprisoned themselves in order that they might be more free." At the end of *Moradas*, her last major piece of writing, Teresa describes her spiritual interior, her soul, as a cloister in its own right. The soul was a roomy place with concentric corridors and hidden chambers, and God dwelled at its center. Addressing her sisters in the Carmelite order, she wrote, "I think you will find consolation and delight in this interior castle, where without having to ask permission from your superiors you will be able to go inside and walk around wherever you want."

In midcareer Teresa became a reformer of the Carmelites. Inner reform initially, as she resolved to live in holy poverty. Chafed by rough wool habits, walking barefoot or in thin hemp sandals, she and her sisters survived on alms. They plotted against the needs of their bodies and flogged themselves in penance if their bodies complained. As much as feasible, they cut their contact with the outside world. The paradoxical responsibility of the Discalced (Shoeless) Carmelites was to shine the light of their chosenness on all of mankind, yet this was a duty they could perform indirectly, by obeying the rules of their insular order.

Next, external reform. The idea of asceticism grew strong in Spain, as it periodically does in the history of the Catholic Church. (Saint

Cajetan went down the same path a little earlier in Italy.) From her spiritual garrison in Avila, Teresa made forays into Castile to establish new convents. Her second convent was at Medina, about fifty miles away. Most cities preferred nuns of Old Christian stock, but Medina was sympathetic to *conversos*. Teresa went door-to-door throughout the province, gathering housing and support for her austere foundations. A Franciscan missionary who was just back from saving souls in the Americas encouraged her plan.

In founding her string of missions Teresa turned out to be a good politician, which may be half the battle to becoming a saint. She was by turns very practical and very mystical. When her body was writhing, even levitating, it was said, during her rapturous encounters with God, she was always careful, afterward, to minimize the experiences. She would publicly doubt her ecstasies, worried that Satan might have tricked her while she was entranced. For it was a trap to fall too easily into love with God; besides, the Inquisition might challenge it. During Teresa's most celebrated rapture, rendered later in painting and sculpture, a cherub or seraph plunges a red-hot spear into her heart over and over—the sweetest pain, as she describes it, penetrating to her entrails and leaving her fired with love for God. Over the top as this may be, Teresa had a brisk, clinical way with words, as if she were reporting an event quite separate from herself. If a mystic is a person who develops a private route to the divine and wishes to draw a map that people can follow, she must have an unusually rational mind.

The comparison that's being drawn here between Teresa and Shonnie Medina will not stand heavy scrutiny, admittedly. Pure believer cloistered in her mountain valley, superstitious *converso*, creature of her windowless Kingdom Hall, Shonnie left no record of her spiritual development. She was not a nun, nor did her devotion lead her to imitate Christ. The arrow that pierced her chest came not from God but from Lady Death's bow, and was aimed not at her soul but at its analog, DNA. For DNA is science's answer to the soul, is it not? Liberal theologians have toyed with the notion that DNA is the missing link between the body and the soul, which have been held separate in Christian thought since Teresa's time. DNA both inhabits the flesh and informs the spirit. By the same token you could say that the goal of Teresa's ardent, analytical writing was to construct a phenotype of the soul.

The Jehovah's Witnesses do not believe that the soul is independent or can be cut loose from the body, or can be parsed in any way. Soul, they think, is just another term for life or organism. Man does not *have* a soul—he *is* a soul. This squares with the concept of soul in the Hebrew Bible, the body and its animating spirit as one. Set the soul aside then. It is through the suffering of two imperfect bodies, Teresa's and Shonnie's, that the most fruitful comparisons may be drawn.

Shortly after deciding to become a nun, the young Teresa fell gravely ill and was bedridden and paralyzed for nearly three years. The learned doctors couldn't help, nor could a folk-medicine practitioner whom her family consulted. Regaining her strength, Teresa was bothered for the rest of her life by nonspecific symptoms, an array of pains and nauseas and noises in the head. These seemed to synchronize with her accomplishments, so that, as one of her biographers remarks, "[T]he busier she became, the more her infirmities got in her way."

It's easy to assign a psychosomatic label to Teresa. The caveat is that her symptoms caused her spirit to rejoice. She exceeded Shonnie in equanimity to pain. She not only worked her discomforts into her normal plan of bodily mortification, but also applied them to her unscheduled spiritual transports, the helpless paralysis and pricks delivered by God. What made her miserable was the soul's torment, its moments of what she called dryness or doubt, and not her body's torments. Death was her desire. During her raptures Teresa felt her body deliciously close to death. When, exhausted and bleeding internally, she truly came to the end, she lingered by the exit and orchestrated her departure with flourishes. "Let me suffer or let me die" remains her most famous saying.

The Spanish Catholic disgust or disillusionment with the human body can be traced to Saint Paul, who urged Christians to dispense with the deeds of the flesh and to mortify the flesh in order to live. But such is the hold of superstition, down in the trenches of belief, that after Teresa's passing, parts of her body were chopped into holy relics and distributed to the rich and powerful. Physical pieces of Teresa held sway in Spain centuries after her soul had left.

Jehovah's Witnesses believe they will retain their bodies forever. The body is the dominant entity when they contemplate the afterlife. Although from birth the body is beset with flaws and disease, the believer will get it back shiny and restored in the Earthly Paradise to come. More than other Christians, the Witnesses insist on a physical resurrection. The body's resurrection—theirs but probably not yours—will take place after the Apocalypse. It's all laid out in the Bible, if you can read the Bible properly.

Shonnie did not yearn to die. Shot through the breast, her treatments come to naught, she was scared as she neared the end. But all her visitors attest to her composure when they would find her sitting up in bed, her long hair brushed and her nail polish without a chip. She went out in style and was never tempted, as Job was by his wife, to curse God and die.

We are imperfect, said her sister, Iona. It's not normal to be sick and to die. At times she was freaking out, but she never was completely lost. She made the best of it. We talked about it a lot. It's human imperfection, her fear. Her attitude was, I am afraid but Jehovah is going to fix it. We grew up living the way Jehovah wants us to. She was not helpless.

Sepharad—what became of it? O the lamentation as the Sephardim were banished and drifted away from Iberia. Some went north to Amsterdam, some west to Mexico, some even into Muslim Palestine. It was in Palestine that a rabbi named Isaac Luria, a contemporary of Teresa's, helped to shape the Kabbalistic school of Judaism. According to the mystical Kabbalah, the Jewish soul has all sorts of options independent of the body.

In the Kabbalah there is a concept called *ibbur*, which describes one righteous soul possessing another. *Ibbur* means impregnation or incubation. It is a positive thing, not like being taken over by Satan. For a while, two linked souls might dwell in a single body, Teresa tending to Shonnie.

Not steeling the young woman for death, necessarily, but entering her interior castle long before then and comforting her after the little girl was sexually molested by an uncle. The chink in her castle that Shonnie never talked about until she married.

Lucubration, look the night is breaking. . . . The coyotes that were crying have stopped their crying. It's dawn in San Luis, the hub of Culebra.

Chapter Five

THE LOST TRIBE

Marianne Medina seemed to grow closer to Earth as she aged. Phlegmatic where she once had been fiery, she moved through rooms with barely a rustle. The sense of quiet and moderation about Marianne was distinct from the occasional heaviness or sadness she felt over the people she had lost. The oval line about her face rounded, and she went gray at the temples. Marianne Medina was a strong woman, a strong woman affected by gravity.

In early 2007, she and her husband, Joseph, were getting ready to turn over their house in Culebra to Iona, their remaining daughter. Iona and her husband would move in, and Joseph and Marianne, generous parents, would live in a new building that Joseph was constructing behind the restaurant. The move wasn't far, just down the road. Although Marianne had set no firm date for leaving, she was filling boxes in her living room and taking pictures down from the walls. She apologized for the mild disarray of the house.

Shonnie's old room, at the top of the stairs, hadn't been touched. The room had not changed since Shonnie had vacated it fifteen years earlier, following her wedding. (Shonnie did come home several times in 1998 when she was ill.) Thanks to a skylight, the ambience in the small room was cheerful. There was a twin bed with a quilt; a bureau;

a tidy closet. Prints with Indian themes hung on the wall. Marianne's gaze moved about the room palpating for memories, which then issued as stories. How Shonnie helped to make the wedding dress. How she liked to ride Hot Smoke in the pasture and up onto *la sierra*. Her modeling. How she earned Pioneer status going door-to-door.

Her full name was Shonnie Christine Medina, Marianne said. The spelling was supposed to be Shon-i, that was the Indian spelling we wanted, but it came out Shonnie on the birth certificate. Marianne shrugged slightly, as if to indicate that this sort of thing commonly happened to native people and also that her daughter didn't care. There was no special reason for Christine, she added. It just kind of goes with Shonnie.

Opening a bureau drawer, Marianne took out a high school yearbook, some newspaper clippings, and a packet of old photos and spread them on the bed. By now she was smiling and tickled in spite of herself, the sparkle of her scintillating daughter having rubbed off on her, the way that someone else's sequins can't help but make you smile. That Shonnie—she was so vain, Marianne said. She wouldn't walk out the door if she had one pimple on her face. She thought she had an acne problem—she didn't. Her daughters' remedy for pimples was to dry them out with dabs of toothpaste, though no one except Marianne and Iona ever saw Shonnie Medina with toothpaste on her face. Not the pizza delivery boys in Alamosa, who would notice when an order was phoned in from her house. Some of them would drive out of their way, into another guy's territory, just to get a look at Shonnie when she opened the door.

Shonnie was always emotional, Marianne went on rapidly. She'd be happy—too happy—but then a down day would follow. Marianne blamed Shonnie's mood swings on her childhood secret of having been molested. The story eventually came out. A teenage uncle, one of Joseph's brothers, had been an occasional babysitter. It didn't involve penetration, Marianne said, but it was contact and it happened more than once. Marianne was surprised because, as she put it, My family is harsh that way but Joe's is not. Shy and sly, Marianne said of her brother-in-law, a bitter edge to her voice. He traumatized her. Shonnie was sensitive and holding all this stuff in.

It could explain her presentiment about death, her mother thought. If you get too happy, Shonnie had warned, things will happen to you. That was her attitude, Marianne said. When I told her once that she looked like her grandmother, Shonnie said, That means I'll die young.

Your mother?

My mother. You know how they are—she ignored a uterine problem. She died of uterine cancer when she was fifty-three. She treated herself with herbs. Yet Marianne Medina practiced natural medicine too. When her kids had coughs or ear infections, she would take them to a *médica*, a female folk healer, in Taos rather than to a doctor for antibiotics. Marianne was even tougher about her own care, on one occasion yanking out a bad tooth with a pair of pliers.

The photographs that Marianne kept in Shonnie's drawer were taken in the summer of 1997. George Casias, a professional photographer and family friend, had known Shonnie since she was a girl. He had photographed her wedding. When he would ask her to pose, she would always be willing. This time George wanted to try out a new brand of Kodak film. With her anniversary coming up, Shonnie hoped to surprise Michael with some nice shots. She must also have welcomed the distraction from the tumor in her breast, which had been diagnosed about that time.

Shonnie dressed up as an Apache Indian. She wrapped herself in a serape, a full-body garment, striped and colorful. She pinned a plumed ornament, called a scalp tie, to her long, black hair, and brushed her hair over to one side of her face. For jewelry she wore a heavy white necklace, triple-stranded and clasped with a medallion of beaten silver. In the pictures Shonnie is kneeling, her arms and legs folded inside the serape, while her back rests against the furrowed bark of a large cottonwood tree. Two gray-and-white puppies, which look to be huskies, nestle by the trunk. Having these dogs in the photos added a fine native touch, because until the Apaches obtained horses from the Spaniards, they would use dogs to move their possessions over the prairie. The Indian travois, a simple rig of lashed-together poles, was dragged by dogs. The people walked; only if she were very sick would a person go by travois. The vertical stripes of the serape, the deep grooves of the cottonwood tree, the pricked-up ears of the puppies are in sharp focus, but for some reason, maybe because loose strands of hair are blowing in front of her eyes, Shonnie's features seem soft, and her expression seems distant, as if she were looking through the lens at a point far away.

Still, it was dressing up; Shonnie didn't have any investment in being an Indian except at this moment. She was a young woman having her picture taken, one who almost always liked having her picture taken. Marianne Medina definitely did *not* like to be photographed, as the family's video and photo albums make clear. It is comical to watch Marianne slide out of the frame or scurry behind her daughters when the camera attempts to catch her. I just don't like how I look, she said, in spite of having worked for a time as a department-store model in Denver. But more than that, she evinced the deep-seated unease that native people have about photography, the intuition that each shutter-snap steals a portion of their selves. It takes away your spirit, she recalled her father saying.

Marianne's go-to identity, herself in her heart, was Native American. Her mother, Rose, was a Pueblo Indian. Rose's family came from Picuris Pueblo, Marianne said, Picuris and Taos representing the northernmost villages of New Mexico's Pueblo tribes. Juan Quintana, Marianne's father, was Jicarilla [hic-a-REH-a] Apache. But both of her parents also had Spanish blood. Since most of the Indians of northern New Mexico and southern Colorado had Hispano names and ancestry, it was up to them to decide which part of their heritage to emphasize. Juan Quintana thought of himself as an American Indian, she said. His parents and grandparents had lived on the Jicarilla reservation, west of the Valley. He had prominent gray eyes. He was a well-built Jicarilla Apache, Marianne remembered. He was *genízaro*—do you know what that means? It means we used to be slaves.

Marianne Quintana grew up on the wrong side of the tracks in Alamosa, then and now the Anglo center of the San Luis Valley. Alamosa is
about fifty miles from Culebra, on the Rio Grande River, west of Mount
Blanca. When she was a schoolgirl, her ill-defined race was used against
her. They called me Mexican . . . or Chinese, she said. The teachers, who
were white, made kids like me wear gloves when it was our turn to give
out snacks to the class. Get your greasy hair off the desk, they said. We
had to sit in the back of the classroom. We even had separate chalk-
boards and erasers.

In her teens Marianne rebelled. I called a teacher white trash once
after she called me Mexican, and she slapped me. In high school I was
in AIM [the American Indian Movement] and I was also in the Chi-
cano movement. I wore combat boots and fatigues all the time. I was
violent and aggressive. I broke a girl's eye once, and another time I hung
a girl on a telephone pole. You know the rungs they use to climb telephone poles? I pushed her up against the telephone pole and her coat
got hooked up on the rung. All that anger came from what happened to
my dad. Because they stripped him of his culture.

What would have become of Marianne, a great-looking package of
trouble, if she hadn't met Joseph Medina? He was a scrapper too, but
he was secure in his Hispano identity, a proud son of Culebra. More
important, race never was an issue with him. One night, while playing with his band at a high school dance in Alamosa, he looked down
from the stage and saw her. So Joseph married Marianne when she was
seventeen and treated her soothingly, like a high-strung pony. For years
she did not get over her hostility to whites and her defensiveness about
being an Indian. *Indita,* her father-in-law would hiss at her when he got
annoyed. Becoming a Jehovah's Witness helped, but even after becoming a Witness she would get in your face if you crossed her. Her calming
had mainly to do with time and inertia. As the cares of her life pulled her
closer to Earth, the earth pruned away feelings that were unnecessary.

Her oldest daughter, fiery in her own way, could put on an Indian
identity much more easily than Marianne was able to let go of hers.

Shonnie mixed-and-matched—she had the plasticity of a sunflower. She was the Hispano once-removed. She was the middle-class spouse skiing with her Anglo husband, the ardent and prim Jehovah's Witness, the beauty queen and photo model, the cowgirl on her horse, the breast-cancer patient bucking the medical system. For fun, Shonnie was an Apache. But Marianne, even in her lightest moments, did not stray far from her centripetal self.

One last thing, Marianne Medina loved to dance. Pivoting from her solid center, she would spin her grinning husband around the floor. You could take a picture of Marianne when she was dancing and she wouldn't mind. She danced with Shonnie and Iona when the girls were growing up; she organized a dance troupe consisting of her children and four of their friends. Brightly turned out in hand-sewn dresses, the group performed at weddings, doing the *flamenco, mariposa,* and other *folklorico* steps, six little ladies in a nervous line on the shiny floor. Marianne had them change their dresses and ruffled petticoats in the middle of the set. The last time you see Shonnie on videotape, the young wife and her mother are dancing the *cumbia* together, their hands not touching and their steps synchronized. They're as dreamy-eyed as two lovers, while Michael slouches against a wall.

Joy in dancing might spring from the Spanish part of their legacy, overriding the stolid Native American component. No, that's lazy thinking. Stolid is a racial stereotype, so too the portrait of the poker-faced Indian, as if their DNA contained a gene for impassiveness. Indians were said by white people to be fatalistic. This quality of theirs was a defense mechanism, a cultural artifact, and technically not even a Native American artifact. It was a trait that Europeans had induced in the other race and mistook for biology.

Indians lived by what might be called the rule of natural contingency. What was going to happen? It depends. See that rock on the horizon? It could embody a benevolent ancestor, or a witch, or it could conceal an all-too-real enemy who had a knife. The native constantly had to adapt himself to changes in his circumstances. To the Indian, augury was everywhere and the environment was all. Having little control over things, he floated on the swift currents of the world, constantly adjusting his course lest he be dashed and broken on the rocks. Does that make for "passivity" or "fatalism"? No outcome was absolute or essential. No authority could overcome the influence of the landscape or the sway of Earth. In order to survive, every village, every encampment, every campfire installed its own theocracy.

Their fatalism, by which is meant the cliché of the fatalistic Indian, is analogous to the mendelian view of genetics. The mendelian model (from Mendel) oversimplifies genetics by overemphasizing the power of the single gene. The contingent, shape-shifting universe of the Indian was truer to the world of modern genomics, where each person's DNA and environment interacted to form traits unique to that person. In that sense Shonnie was a more natural, more resilient Indian than Marianne because Shonnie wasn't hung up on her identity. Shonnie adapted. At the same time she carried a big-time mendelian gene.

I'm light-skinned, Marianne was saying. I am not a Ute [Ute Indians were another important people of San Luis Valley] because Ute means dark. But she allowed that her pale, almost porcelain coloring might stem from a rival source, a Chinese cook who had worked with her mother. The Indian man Marianne regarded as her father, Juan Quintana, who helped to raise her and provided her name, had never married her mother. Was my biological father Chinese? she wondered, becoming animated again. Yes, maybe. His name was Pee Wee Wong. I remember he gave me a Chinese doll. Of her mother's dozen children, Marianne was the only one whose paternity was in doubt.

Was she aware that DNA analysis could answer her question? Harry Ostrer happened to have a contact at 23andMe, a gene-testing company, and he arranged for Marianne to get a free test kit. If she would swab cells from the inside of her cheek and send in the sample, the company would tell her if her DNA contained the genetic markers of Asians (i.e., Pee Wee Wong). But as of this writing Marianne hasn't wanted to know more about her biological father. Apparently she wants to maintain the identity she has now.

———

After you have approached the San Luis Valley a few times, whether through the snowy ranges on the north, east, or west, or from the broad Taos Plateau to the south, you appreciate the centrality of Mount Blanca. The mountain is a psychological force in the Valley even more than a physical anchor. Nothing happens in Culebra that Blanca doesn't supervise.

What the Navajo Indians called White Shell Mountain marked the eastern edge of their homeland. As long as they could see Blanca, they felt secure. The Navajo and other roaming Indian people believed that the mountain stood at the doorway to the subterranean spirit world. As noted, Blanca served as the midwife for the emergence of the first human beings.

Geologists, who are a recent tribe of human beings, say that the earth cracked and pulled apart here, and that the mountain and the rest of the Sangre de Cristos heaved up along the faults. A long fissure or rift running south from Blanca became the birth canal of New Mexico because it housed the Rio Grande River, against whose flow Europeans forged their way north, bearing their DNA to the Indians.

For geneticists, the story of the creation of New Mexico and New Mexicans is straightforward: Spaniards mated with Native Americans, the result a hybrid people called Hispanos. The snobbish distinction that Marianne's father-in-law, Joe U. Medina, maintained—that there was a gulf between his people of Spanish ancestry and Native Americans—does not hold water on the level of DNA. A 2009 genetic analysis of Hispanic college students in New Mexico found that the study participants regularly underestimated the proportion of their Native American blood. For centuries Hispanos have told themselves and the outside world they were Spanish; time and again Spanish was the face put forward by the local historians. Since history is written by the victors, as Winston Churchill said, the founding of New Mexico starred bold Spanish conquistadors and toiling Franciscan friars.

Watching a Western they never grew tired of, Hispanos saw themselves pass from noble Spanish origins through a brief phase of Mexican rule to a shotgun marriage with Anglo Americans, which turned out happily by the time the credits rolled. The mutely receptive setting, the spare and sunlit stage, was the Indians'.

Historians in the modern, or postmodern, period have demurred. The assumption is that all the conventional narratives have to be reexamined because they suppress facts and embody prejudices about minority cultures. Thus the edgily titled *When Jesus Came, the Corn Mothers Went Away: Marriage, Sexuality, and Power in New Mexico, 1500–1846*, by Ramon Gutiérrez (1991), a vital work of revisionist scholarship stressing the Indians' role. For another example, Estevan Rael-Gálvez, recently the state historian of New Mexico, planned a book that he called *The Silence of Slavery: Narratives of American Indian and Mexican Servitude and its Legacy*. Rael-Gálvez is proud to have uncovered Native Americans in his family tree. He probably speaks for the majority of historians when he says that the story of the United States is that cultural strains are erased, and that indigenous strains are erased most of all.

The story begins in 1598, when an expedition of several hundred Spanish colonists led by Juan de Oñate crossed the Rio Grande near present-day El Paso. Most were unmarried young soldiers. Some of the officers and civilians had brought along their wives and children. Spearheading the *entrada* was a tatterdemalion group of friars in gray robes. The first Franciscan missionaries to New Mexico emulated the twelve apostles with a self-conscious but deeply felt zeal. They said a Mass in the desert and thanked God for the new land stretching before them. Unlike earlier conquests in the Americas, whose object was treasure, the northern extension of New Spain was intended to be peaceful and evangelical. Gaspar Pérez de Villagrá, an officer of the expedition who wrote the first history of New Mexico, found it significant that El Paso and Jerusalem lay on the same latitude.

In this New Jerusalem of the Rio Grande del Norte lived forty-five thousand Pueblo Indians, by modern estimates. (*Pueblo* means town or village in Spanish.) These first natives to encounter the Spaniards lived in well-developed societies along the river. The Pueblo people hunted, fished, and farmed. The bottomland of the river provided good soil for growing corn, their main crop, and also beans, squash, cotton, and sunflowers. The outsiders would have recognized the sunflower. In that era the well-traveled plants were being cultivated both in Mexico and in Spain. The Indians made use of the sunflower seeds for oil and medicine, the petals for dye, and they set their hunting calendars according to the sunflowers' growth.

Pueblos looked something like rectangular Lego constructions, with interlocking apartments and multiple stories. You entered a house not through the ground floor but by a ladder, which could be pulled up afterward for safety. Women were in charge of the houses, while men focused on the *kiva*, the religious center of the village. The *kiva* was circular in shape, and it too lacked a ground-level entrance, being accessed through a hole in the roof. Along its walls were altars with various animal fetishes. The floor of the *kiva* contained a navel, meant to be a passageway for spiritual traffic in and out of the earth.

The Indians' landscape was long not only in horizontal dimension but also in vertical thrust. Pueblo women were stationed closer to the earth than men, because they bore fruit like the earth. The domain of males was airborne—in the realm of clouds, lightning, and rain, the tempestuous things that fertilized the earth. The Indian deities, collectively called *katsinas*, dwelled in the rain clouds, as did the departed ancestors. In their cosmos serpents (*culebra* to the Spaniards) corresponded to penises, and the snake dance to summon life-giving rain began with men slitting their penises and bleeding onto the dry ground, seeking a religious ecstasy in the pain.

The Spaniards told the Puebloans they must surrender all of these beliefs. Although the soldiers and friars had come in peace, refusal to convert to Catholicism was not an option. The same determination that had forced Jews and Muslims to knuckle under in Iberia was applied to the Indians, who were (it appeared) a heathenish people in the grip of Satan and a manifestly inferior race. Nonetheless, they were Spanish subjects and must comply. What's more, they must supply food for the Spaniards and unpaid labor for the building of the missions and the *conventos* where the friars would live. Quite so, and for the next eighty years the New Mexico pioneers sustained themselves exactly as a parasite does, by affixing themselves to the pueblos and sucking nourishment through the Indian walls.

The San Luis Valley is treated as something of a footnote in the settlement of New Mexico because the Spaniards never occupied it. Sequestered by mountains, the Valley was a no-man's-land between the sedentary Pueblo tribes and the nomadic or Plains Indians, who were Navajos, Utes, Jicarilla Apaches, and Comanches. It took much longer for the tribes on the menacing northern fringe of the colony to be brought into the genetic mix and erased.

An early commentator, the American Josiah Gregg, in his 1845 account of the Santa Fe trade, tersely summarized the Spaniards' relationship with the Indians, "upon whom they forced baptism and the cross in exchange for the vast possessions of which they robbed them." When one of the western pueblos, Acoma, balked at the arrangement, Juan de Oñate showed his conquistador's teeth. Besieging Acoma and defeating the Indians, Oñate ordered that every able-bodied man who was left in the village should have his foot cut off as a lesson to the other pueblos. Acoma's children then were distributed to the colonists as slaves.

The Franciscan friars complained about Oñate's methods, and within a few years he was replaced as governor. Church and state quarreled often in the early years of the colony. For all of their sacerdotal superiority, the friars tried to be kind to the Indians. When the Puebloans brought gifts or donated labor, the brothers reciprocated with livestock, seeds, and clothing—as much as they had, they tried to give away. In the manner of Saint Francis they had sworn themselves to selfless poverty, although, unlike Francis, the friars used their material goods to shape the Indians' behavior and called on the military to enforce their will. The Indians were encouraged, sternly encouraged, to kiss the friars' feet.

The Franciscans were experienced at spiritual warfare from having dealt with native peoples in Mexico. The *padres* took credit for rain-making and healing and hunting success—the powers the people ascribed to the *katsina* spirits. They superimposed their adobe missions on the Indian *kiva* sites and substituted their icons and relics for the animal fetishes on the Indians' altars. Because by happy coincidence the Christian cross resembled the Pueblo prayer-stick, the friars made sure to enter an unfamiliar village brandishing their most important symbol. The Catholic saints were arrayed against the pantheon of the *katsinas*, and the paramount figure of the Virgin Mary, mother of Jesus, supplanted the life-giving Corn Mothers while clothing herself in some of the Corn Mothers' imagery. Holy days were adjusted so that they fell on Indian feast days. And when the missionaries found out that Pueblo warriors whipped their bodies with cactus in order to toughen up for battle, the Franciscans were pleased to demonstrate their own mortifications, dragging huge crosses through the pueblos, with blood dripping from their bare, striped backs.

So the Christianity that was forged in the Kingdom of New Mexico, and that endures today in little corners of Culebra, was part Indian. A flamboyant, demon-riddled Catholicism from sixteenth-century Spain found a mate in the out-of-body expressionism of Native Americans. Syncretism is what you call it, syncretism meaning religious combination or religious admixture. That time when Marianne snatched the girls from the *morada* because a ghoul had scared them, a man who was wearing a skeleton outfit? Well, he had jumped out of a trap door in the floor, as if he were entering the world through Earth's navel. The *morada* of the *penitentes* descends partially from the Indians' sacred *kiva*. The *morada* is a sort of *katsina* clubhouse in Culebra.

Now as to the syncretism of bodies and blood. During the seventeenth century, there were dozens of Franciscans in New Mexico, who kept their hands off the Indians so far as is known, but there were hundreds of settlers and soldiers who were unrestrained. The conditions around the pueblos might be described as a sexual free-for-all. Uninhibited, the Indians copulated frequently and openly, and women offered their bodies as gifts to the strangers, per the Pueblo custom. The Spanish men naturally accepted, thinking the husbands either dishonorable or dupes. As the settlers began to demand tributes from the Indians and to acquire their children as household slaves, the sexual exploitation found many new avenues. The *padres* wrung their hands but did not or could not stop it, for they were more concerned with reforming the Indians than their own countrymen. Very soon the mixed-race children, the Hispano *mestizos*, appeared.

Thunderstorm and lightning, the sky ejaculating, the airborne ancestors writhing in dark tumult—the kind of weather you might experience at the base of Mount Blanca on a summer afternoon. Upsweeping wind, sudden spatter on the windshield. In the pauses of the storm there comes virga, wispy tendrils of precipitation. Virga doesn't touch the ground but wicks moisture from the clouds like the fringe on a wet buckskin jacket.

Early the next morning, when the air is washed and still, Blanca puts on terrycloth clouds, which slip to her midsection as the morning progresses.

The *mestizaje*, the racial mixing process, had started even before the conquest of New Mexico. By dint of breeding with the native tribes of Central America, many of the Spaniards already had some Indian blood, and the blood of African slaves had entered their gene pool as well. The move up the Rio Grande brought a major new infusion of Indian DNA.

More comprehensive than any historical document, DNA writes a record of matings in its four-letter language, although the record isn't arranged sequentially or chronologically like a family tree. It's a jumble, like a drawer stuffed full of parking tickets or grocery receipts with the dates snipped off. In each new generation of human beings, the identifying marks of the previous generations are halved and shuffled on the chromosomes to make room for additional markers. But since humans tend to marry their own kind, the same variants of DNA are shuffled in and out. When an admixture takes place, an intense crossing-over between two peoples or races, the event makes a large impression on the DNA. It can be seen long afterward unless it is obscured by more recent admixture.

A 2004 study showed that the Hispanos in San Luis Valley are about one-third Indian and two-thirds Spanish-European. They have a small portion of African ancestry, averaging 3 percent. The Hispanos generally resemble other Hispanic and Mexican-American groups, while having a somewhat higher proportion of European blood than the rest. Genetic research also has confirmed the harshly one-sided nature of the admixture. By paying special attention to the Y chromosome and the mitochondrial DNA (mtDNA), scientists proved that the genetic exchange in the early years of New Mexico was almost entirely between Spanish males and Indian females.

The Y chromosome of males, handed down from father to son, was discussed previously; the mitochondrial DNA is a small, separate stash of genes within the cell but outside the nucleus and chromosomes. Inherited through the maternal line with no input from fathers, the mitochondrial DNA provides a narrow but relatively unbroken view of female ancestry. It's your mother's ancestry as seen through the floor of a *kiva*, just as the Y chromosome narrowly reveals a part of a man's paternity. The Y chromosome of Hispano men is hardly Native American at all, while their mtDNA is about 85 percent Indian. Again, the former represents fatherhood, the latter motherhood. The skew between the two means that mating happened in one direction. It means that Indian men and Spanish women were largely on the sidelines when the admixture between Spanish men and Indian women occurred. Indeed, throughout Central and South America the same DNA pattern is found—the echo of the Big Bang at the formation of the Hispanic universe.

The intercourse that turned Spaniards into New Mexicans continued for decades, geneticists believe, extending from the Pueblo tribes to the more resistant blood of the Navajos, Apaches, and Utes. But after the Hispanos were formed, mating took place within a closed circle. Europe would send no more of its genes. The historical record indicates that the Kingdom of New Mexico had very little immigration after being established. In fact, many of its colonists left, discouraged, as Juan de Oñate was, by the lack of mineral wealth and by the hostile tribes beyond the pueblos.

The Puebloans converted to the Catholic faith en masse, or they claimed to, and their children, the *mestizo* cohort, were ready-made believers. If their fathers acknowledged them as sons and daughters and took them in, they were deemed *españole* as well. This wasn't how the Franciscan brothers had imagined their New Jerusalem would go—pagans won over to Christ through a change in their genes rather than in their hearts. Their hearts, in fact, were much slower to change.

In 1680, the Pueblo tribes of New Mexico revolted. After years of subjugation and months of planning, the Indians rose up simultaneously. Although their numbers had been much reduced by smallpox, a disease unleashed by the Europeans, and by forced relocations of their villages, still the Indians outnumbered the Spaniards and their dependents by ten to one. (The genetic imbalance was less extreme, but in the dichotomous culture of the occupied pueblos, people grouped themselves with one side or the other.)

A medicine man named Popé led the insurrection. He had learned the black arts of his people illicitly—how to heal with herbs and exorcisms, how to sprinkle eagle's blood or bear's blood on secret fetishes. Under the noses of the *padres*, the medicine man abstained from eating and from sex, growing gaunt and pure, to augment his power. But in 1675, Popé and dozens of other shamans had been arrested in a colony-wide sweep. Three of the Indians were put to death, the rest whipped for practicing heresy and sorcery. Popé decided that enough was enough. He foresaw an apocalypse.

To get out of reach of the authorities, Popé moved to Taos Pueblo, in the far north. Channeling the *katsina* gods, he prophesied a golden age of abundance, an end to the Indians' droughts and misery. Soon the people would return to the fruitful conditions that had prevailed at the beginning of time, but first the Christians and their God must be driven from the land. By communicating in code, the pueblos managed to keep the revolt secret until nearly the last minute. Over several bloody days they killed four hundred Spaniards and *mestizos*, including twenty-one of the thirty-three Franciscan friars who worked in the province. As the Indians torched the missions and desecrated the Catholic icons with feces, the surviving contingent of two thousand Hispanos fled Santa Fe and retreated to the south.

For thirteen years the Hispano remnant lived in exile in northern Mexico. Then Spain decided to take back New Mexico, the Indians hardly resisting, for Popé's earthly paradise had not materialized in the interim. Some of the pueblos were even relieved, so fearful had they been of retribution. They swore allegiance to Jesus and got back into their traces. The Hispanos after all were not total strangers but blood relatives. New colonists from Mexico supplemented the former colonists, and things went on as before. The usual way that historians treat the Pueblo Revolt is as an asterisk, or a watershed dividing seventeenth-century New Mexico from the more complex, eighteenth-century phase of the society. Continuity is stressed, but still the Pueblo Revolt was the most successful native uprising in American history, Little Big Horn plus Wounded Knee times ten.

The Kingdom of New Mexico was smaller than previously because the western pueblos, such as Zuni and Hopi, were let go after the reconquest. Cautiously, the Hispano *pobladores*, the settlers, spread out from the Rio Grande and established their *ranchos*. They grazed sheep and cattle on the narrow floodplains in winter and moved their animals to the cool, upland benches in the summertime. The Pueblo Indians handled most of the farming and did the menial work. The Hispano villages, called *plazas*, consisted of small adobe dwellings built side by side like a strip of motel rooms. The typical *plaza* had two L-shaped banks of houses facing each other across a square. The enclosed area, or *patio*, served as the communal space of the village, while the outside walls of the housing blocks were left blank and windowless.

The design was all about defense. During the eighteenth century the free-riding Plains Indians threatened the pueblos and the *plazas* unceasingly. Navajos, Apaches, and Utes had acquired horses ("magic dogs" to the Utes) from the Spaniards. The *indios bárbaros*, as they were called, drew strength and poise from being in motion, whereas the stationary pueblo tribes were as tippy as a bicycle at rest. The colony was almost entirely surrounded by a fluid and hostile territory known as *Apachería* territory. Taos Pueblo got too dangerous for the Spanish to occupy, let alone the San Luis Valley farther north. Marauding, slave-raiding, slave-trading, ransom, and rape were practiced by both sides. According to one historian, the Hispanos were judged to be better at stealing people, the Navajos better at stealing livestock and corn.

At the insistence of the Franciscans, slavery of captured Indians was illegal, but not enforced bondage for the purpose of their conversion and improvement. Indian women and children, stolen or bartered, were known as *genízaro* and were as good as slaves. Marianne Medina had said that her father, Juan Quintana, came from *genízaro* Apache stock. They were considered low-class people, lower than the Pueblo Indians. The second- and third-generation *genízaro*, having been stripped of their tribal identities, went off and formed their own villages in the borderland between *Apachería* and New Mexico, where they worshipped Christ in weird and wonderful ways.

Internally the Hispano colony was preoccupied with race, since racial distinctions determined the social order. New Mexico had a *casta* system: *casta* means a person of mixed blood. In addition to *españoles* and *indios*, new categories of citizens had emerged in Santa Fe. The *españole* still was at the top, whether or not he was fully Spanish. A *mestizo* was the child of an *españole* and an Indian woman; a *mulato* of an *españole* and a black woman. A *coyote* was the offspring of a *mestizo* and an Indian. A *lobo* was a cross between a Native American and a Negro. Just as in the animal kingdom a coyote is less dangerous than a wolf, the human *coyote* signified less trouble to the social order than the *lobo*, thanks to the Spanish blood moderating the former. The *genízaro* represented yet another native type, as mentioned. If you weren't sure of the mix in the person you were dealing with, you might refer to him or her as *color quebrado* (broken color).

All this was pre-Mendel, pregenetics. What you saw on the surface of people, whether their clothing or color, was thought to reflect the innate, blood-borne indicators of their race. (In parts of New Spain it was illegal for a non-Indian man to wear Indian clothes.) Fast-forwarding a couple of centuries, scientists have learned that skin pigment is an extremely complex trait. It is controlled by at least one hundred genes in variant forms. The variants act like rheostats, turning up or turning down the amount of melanin in the skin cells. In the grand scheme, evolution, responding to environmental factors such as solar intensity, shapes the genes underlying the differences in human skin color. Although the best predictor of your color is your parents' coloring, the mechanisms of inheritance can cause surprises. The *casta* system was mystified by what were called *torna atrás* (throwbacks). An example would be the child of a white father and a *lobo* woman who turned out blacker than the mother—this due to recessive African and/or Native American variants surfacing in the admixed DNA.

Hispano people as a rule are lighter-skinned than other Hispanics, owing to their higher proportion of Spanish blood. Even within the Hispano population, according to the 2004 study of San Luis Valley DNA, European ancestry is greater in Hispanos of lighter skin. It seems that, among New Mexicans, skin pigment was lightened without waiting for evolution. The reason is something called assortative mating. When whites marry whites, and *mestizos* marry *españoles* rather than *indios*, the children will on average be whiter. In Santa Fe during the eighteenth century a distaste for dark skin led people of all hues to generate kids who were lighter. First cousins from wealthy, white *española* families sought dispensations from the Church to marry each other. The dispensations were granted because of the importance of keeping Spanish blood pure, outweighing the problem of consanguineous marriage. The offspring of such couples were more likely to survive than the darker, out-of-wedlock babies of *genízaro* Indian women, even if the fathers were the same men.

Recall the Spanish ethos of *limpieza de sangre*, blood that was free of foreign taint, blood that had vanquished the Muslims and routed the Jews. In colonial documents and letters the Indians are disparaged as Moors and their *kivas* compared to mosques. As for the Jews, in 1751 a man reported that he had beaten his Indian slave with every right, "so that I might do to these Jews what they did to Our Holiest Lord." The Old Christians of New Mexico feared that the swarthy New Christians could lapse at any moment into a primitive state, as they had in 1680. The sexual mingling taking place on the down-low did not allay these fears but only increased them, as the *españole*, despite his posturing, could feel his essence dribbling away.

All the while, the untamed nomads harassed the colony. During the second half of the eighteenth century Comanche war parties sweeping down from the Great Plains became a powerful enemy, such that the Jicarilla Apaches and the Utes made alliances with New Mexicans against them. In 1779, setting out with eight hundred soldiers and Indian auxiliaries, Governor Juan Bautista de Anza circled through the San Luis Valley and attacked the Comanches from the north. Anza marched his army at night so that Comanche scouts would not spot its dust cloud on the horizon. Besides gaining a major victory, Anza is credited with naming the Valley. San Luis, or Saint Louis, was King Louis IX of France and a Crusader against the Muslims.

By this time, however, Spain had nearly given up on New Mexico, a poor place with few resources and inconstant souls. Almost nothing was being invested there, almost nothing extracted. About the best that could be said of the colony and its thirty thousand inhabitants was that they were self-sufficient. The bishops in Mexico City and Durango stopped sending Spanish priests to serve in the north. Ignorant of events in the rest of the world, the Hispano gentry, *hidalgos* in their own imaginations, like little Quixotes, grazed their flocks, managed their stone-faced servants, and patrolled their baked-mud domains. Fray Angélico

Chavez, New Mexico's best-known historian, equates them during this period to a lost tribe. When Mexico broke away from Spain in 1821, the title to the place changed hands without notice.

The new government ended the *casta* system of racial discrimination, at least officially. Religious standards having been slipping for years, folk Catholicism and Indian rites consorted with each other out in the open. Priests took common-law wives. Not only was the province a backwater, but also the towns within the province were cut off from one another because of the Indian threat. The former regime had restricted travel between villages, partly for public safety, partly so that the pueblos wouldn't plot insurrection. It's estimated that 90 percent of the marriages took place between people who lived in the same town or pueblo. The Hispanos were ghettoized along the Rio Grande. The isolation and inbreeding had genetic consequences that still register today. Even in the twentieth century, because births took place out of wedlock and birth certificates were inaccurate, you might marry a close cousin and not realize it. As was pointed out in chapter 3, when a population becomes separated geographically or culturally, a genetic disease originating in a few persons can affect a large number of descendants, because of the phenomenon of genetic drift.

In effect the Hispanos experienced a bottleneck. Bottleneck usually implies that a group is reduced in size before expanding, but here it merely suggests a small population bottled up in its location. On the one hand, harmful mutations aren't vented and dispersed; they may multiply unpredictably. On the other hand, the persons who inherit the genetic disorders may not survive or, if they do, they tend not to have kids themselves, which puts a brake on the spread of the mutations.

In families from northern New Mexico and southern Colorado, doctors have identified a rare type of muscular dystrophy (OPMD for short) that weakens the eye and throat muscles progressively; cavernous angioma, an inherited malformation of the blood vessels of the brain; a unique strain of dwarfism, as yet unnamed; and recombinant chromosome 8 syndrome, which involves mental retardation and heart problems and also goes by the name San Luis Valley syndrome. In the 1980s, an investigator of San Luis Valley syndrome, working with affected families in different states, traced the DNA disruption back to a common ancestor, the founder, who lived in northern New Mexico in the nineteenth century.

To this unfortunate collection of disorders can now be added a cancer syndrome caused by 185delAG. The gene was very slow to be detected in the Hispano community because it hid behind the background cases of sporadic breast and ovarian cancer. With families reluctant to talk about cancer and physicians not asking patients about relatives with the disease, the mutation multiplied freely. Potentially, the mutation is as pervasive among the Hispano people as among Ashkenazi Jews. The geneticists don't know because they have not studied Hispanos to nearly the same degree, but the gene may well be altering the cancer statistics of the region surreptitiously. Scientists do know that Hispanic and Latina women—a grouping that includes Hispanos—are at lower risk for breast cancer than non-Hispanics, though the risk rises in proportion to the amount of their European admixture.

Genetic disease also infiltrated the pueblos of the uncomplaining Indians. Reduced to twenty villages by the start of the nineteenth century, the Puebloans turned inward. The fragmentation created a string of bottlenecks. One result, at Hopi Pueblo and others, was a relatively high number of albinos. Zuni Pueblo, having expanded to a community of ten thousand, was reported to have the highest measured rate of cystic fibrosis in the world, as well as the highest rate of end-stage kidney disease.

Cystic fibrosis was introduced to the Zunis by a Spanish founder many generations ago. The powerful recessive mutation has a European signature. The DNA variants behind the Zuni kidney-disease problem are unknown but are related to the Indians' high incidence of type 2 diabetes. For the latter two diseases, genes are not the whole story, since the conditions are triggered by dietary and lifestyle changes entering from the American environment. Zuni keeps its suffering from kidney disease private. It is disturbing to visit the tribe's dialysis clinic, which treats three shifts of patients each day. Young bodies as well as the elderly are hooked up to the machines. Their faces are pale and drawn as their mixed blood is cleansed.

When driving from the green corner of Culebra to the scrappy city of Alamosa, first you head north to the forgettable towns of Fort Garland and Blanca, near the foot of the sacred mountain, and then you go west. For miles and miles on a straight road through empty, overgrazed prairie, the mountain never moves.

The noonday August light drains the sagebrush and prickly pear cactus of their coloring. Sunlight saturates the landscape so thoroughly that shadows standing next to objects (a telephone pole, a well casing) are unexpected and alarming, like black holes.

Above the treeline, Blanca's throat is bare and fluted. The rock to the summit is dotted with pearls of snow.

Dorothy Martinez Medina, Shonnie's grandmother, lived in Alamosa in a housing project for low-income seniors. Each of the ranch-style units, with neat sidewalks and squares of lawn between them, contained two or three apartments. With minor rearranging, the project would look like the original New Mexico *plaza*.

The three small rooms of Dorothy's apartment were laid out in a row, so that you could see from the front door through the sitting room, through the kitchen, and into her bedroom. There were lots of things to look at: family photos in frames, stuffed bunnies, brightly colored tchotchkes on the tables and counters, but no religious items. Dorothy's two favorite colors, whether in stripes, checks, or polka dots, seemed to be red and white. Throughout the apartment, pictures of Shonnie showed her at various times of her life. The cutest might be the one of Shonnie as an infant. She is standing up on a table between her two seated parents. It's the early 1970s, and Joseph has long hair and wears a beaded necklace. Marianne's black hair is very long and spills down her back. Joseph was working then as a musician, Marianne as a department store model. A handsome seventies family, just getting started.

Dorothy Medina, the matriarch, *Mamacita* to her kids, was about to turn seventy-five. As of August 2007, she had outlived her ex-husband, seven of her thirteen siblings, four of her fourteen children, and of course her grandchild Shonnie. Five of her sisters had died of breast or ovarian cancer, but Dorothy, though she carried the 185delAG mutation, was unaffected. Possessing a thin, angular body and cheerful, darting eyes, she did not talk much. She looked to be in good health.

She has diabetes and high blood pressure, cautioned her granddaughter Shannon Apodaca, who was managing the meeting today. Shannon worked as a visiting nurse, and she would drop in on her grandmother to make sure Dorothy was taking her pills for her two conditions. Hypertension and high blood sugar are silent ailments that were never diagnosed in Hispanos of earlier generations. But look, said Shannon, she has wrinkle-free skin. Her grandmother beamed. *Mestiza* Dorothy was a little darker than most of her children. Shannon was somewhat dark as well, and heavy-browed and broad-faced like an Indian, but she was a world apart from Culebra, a confident young woman, outspoken and moody. She said, I wish I could be as dedicated as you, Grandma, with my diet.

Dorothy had attended the genetic counseling session for the Medina and Martinez cousins at T-ana's Restaurant. She understood the part about inheritance, but today she wanted to know what genes were. That required a little discussion at the kitchen table, with its red-checked vinyl tablecloth. Shannon asked, Does this stuff about genes scare you? Confiding in a stage voice about her grandmother, She's a worrywart. . . . Aren't you? Nonetheless, a few weeks later Dorothy took the test for 185delAG—her positive result nearly a foregone conclusion—while Shannon, the nurse, put off the testing and continued to put it off, even though, or perhaps because, Chavela, Shannon's mother and Dorothy's daughter, had tested positive for the mutation earlier. Because my mother is a carrier, that means *I* have to be tested, Shannon said, trying to convince herself.

Dorothy Medina straddled two medical camps, the modern one that analyzed her blood sugar and DNA mutations with the aid of government subsidies and the Culebra world of catch-as-catch-can. Every day she drank a cup of her home-brewed herbal tea, half a cup in the morning and half a cup in the evening. Standing before the cupboard, Dorothy pointed to her stores of dried plants. The leaves were collected by her or her knowledgeable friends—women walking in the woods looking for what they could use. She had big mason jars of *chamiso*, yerba buena, manzanilla, wild rose hips (*champe*), and others whose names went by too fast.

In the old days every Hispano village had an amateur doctor or *médico*. The cultural descendant of the Pueblo medicine man, the *médico* would set broken limbs, perform simple surgeries, and treat internal ailments with plants and herbs. As time went on, the healer's role was taken over by a woman, the *médica*, for it was the women who guarded the knowledge of plants and organic processes. The most useful herb for tea was the common prairie shrub *chamiso*, or big sagebrush. Dorothy went to the refrigerator and brought out a small jar. Her children laughingly called it Mama's *whiskito* because she took it everywhere, as other people must bring a flask. The elixir was dark brown. It was very aromatic. In passing from mouth to stomach, the snakelike liquid slithered down rather than swallowed down, a true *culebra* drink, cool and laden with botanical oils. The gullet was left tingling, the mind freshly focused. Dorothy had given the tea to one of her sons when he visited from California recently, and it had relieved his stomachache. *Chamiso* tea had been a staple for Shonnie growing up.

Besides modern medicine and Culebran folk medicine, a third way, a middle approach, was being applied to Dorothy's health. It was modern folk medicine, aka alternative or complementary or New Age medicine. Dorothy kept in the fridge, next to her square jar of tea, a futuristic-looking plastic bottle of FrequenSea, a dietary supplement. Her son-in-law Bill Kramer had recommended it for the treatment of diabetes. FrequenSea's core ingredient was marine phytoplankton, described on its website as a superfood; it also contained extracts of blueberries, aloe vera, frankincense, and rosemary antioxidants, plus plenty of sea minerals and amino acids. FrequenSea was purveyed by a Utah company called ForeverGreen, and Bill sold ForeverGreen health products on commission. Dorothy had no objection to drinking it.

To get to the nub of this digression: Marginal medicine, meaning unproven medical treatments, and marginal religion, meaning the sect of the Jehovah's Witnesses, swirl about the story of Shonnie Medina like two furies. Shonnie inherited a gene, the fifth business of the story, and then two contingent actors came into play, superseding but also springing from her hybrid Hispano background. Shonnie's religion and her medical choices were not connected, not on the face of it, for Jehovah's Witnesses don't espouse alternative therapies as part of their program. Their official advisories about medicine—and about genetics, for that matter—are responsible and fairly accurate. But it does seem that one belief system made her susceptible to the other, just as 185delAG sped the way to her cancer.

Dorothy Medina had converted to the Jehovah's Witnesses in 1983. On this subject the gray-haired, bright-eyed woman suddenly became talkative. It was a treat for her to hold and understand the Bible, she said, and in fact she had learned to read English from the weekly booklets of biblical instruction sent by the Watchtower Society. (The Catholic Church for a long time resisted giving people access to the Bible. Like much else imported from Spain, spiritual truth was hierarchical and dispensed in snippets by the priests.) We don't learn nothing in the Catholic Church, Dorothy exclaimed. They don't read the Bible. They teach us from Jesus and they don't teach us Jehovah. We have to do Jehovah's will, she went on. We have to do what the Bible says— no birthdays, no Christmas, no Mother's Day. She might have added, no icons or *santos*.

Shannon had tried to become a Witness but had lapsed. Five years younger than Shonnie, Shannon had read Bible verses under her cousin's tutelage. I used to study with Shonnie but I quit, she recalled. I was never happier in my life than when I was out in [door-to-door] service. Even after the fourth time I quit, one day I called her again, asking her to study with me. . . . And here she comes! The same day! Shannon remembered with a thrill in her voice.

When Shonnie arrived at her door, Why? Shannon asked her. Why are you doing this for me? Shonnie answered, I'll always come back. Because if I were to die, and I find out that you are not in the new system with me [the paradise coming for Witnesses], I wouldn't be able to stand it.

Shannon deeply missed her cousin, that was obvious. She had learned of Shonnie's disease one day when the two were studying together, when matter-of-factly Shonnie spoke about a lump. Yeah, they told me it was breast cancer, Shonnie said. A year and a half later Shannon went to visit Shonnie in the hospital. Marianne had told her on the phone that everything was OK, and not to bother coming, but then Shonnie called her cousin back and said it wasn't OK, that she was probably going to die.

So I just drove to California and showed up at the hospital, Shannon said. Her hair and nails were beautiful, I remember that. She put me at ease and she was cracking jokes. She said, You ought to see my legs, ooh they're so white. 'Cause Shonnie always had the perfect tan. We started to joke about it. Get some sun, girl, I said.

She asked if I could smell her. I did smell something funny but I said no. Now that I'm a nurse, I know that's the smell of people with cancer. The tumor when it breaks through the skin. They can't smell it anymore themselves.

Shannon's broad face clouded over the tablecloth, like a dark cloud swelling above the checkered fields of Culebra. The *katsinas* in her sky started to cry. Shonnie would be here today, I can't help but think, if she hadn't, if she . . . Shannon gestured weakly, not at her grandmother's potions specifically but at the dramatic decision, beyond any potion, that her cousin had made. I'm *still* angry about it, Shannon concluded, dabbing her eyes and pulling herself together, as Dorothy looked on quietly.

But what was ironic, since small ironies reside in every sad situation, was that Shannon herself had used alternative medicine before having an operation a few years earlier. I had cervical cancer, I wanted it out, she said sheepishly, but just to cover my bases I used my mother's BEFEU [antitoxin] machine, and I went on a Noni [a tropical fruit said to have healing properties] diet. So that was the Native American in her, floating over the rocks of the haunted universe, keeping more than one catechism. Impassive, Dorothy didn't say anything.

Chapter 6

FROM THE *MORADA* TO THE
KINGDOM HALL

A ll that blue sky to fill, and still the clouds line up like pilgrims and process to the tops of the mountains. Narrowly they gather on the Sangre de Cristo range, fold upon fold, white upon white.

You are sitting above the circle of the earth. You are looking down on little Culebra, at the edge of the San Luis Valley. From south to north, the settlements are San Francisco (La Valley), Los Fuertes, San Pablo, Chama, San Pedro, San Luis, and San Acacio. The segmentary landscape is the same as it was when the Hispanos broke ground here, in the 1850s. Long, narrow, parallel lots, called *extensiones*, were set at right angles to the northwesterly flows of the creeks and *acequias* (canals). Every settler's fields lay across the path of the water, and the job of the *mayordomos*, the community-appointed overseers, was to keep the irrigation fair. *Mayordomos* still maintain the *acequias*, but Culebra Creek, after passing through the community, no longer reaches the Rio Grande because of water mining for agriculture in the southern part of the Valley.

Thanks to Google Earth, you can swoop down from the sky on the wintry terrain and can vicariously travel the road between the *morada* in San Francisco and the Kingdom Hall in San Pedro. Golden willow trees,

111

glowing at the branch ends where the new growth is hibernating, make halos of color in the fields. Shrubby red willows line the waterways, their swollen red fingers ready to burst into leaf. Occasionally you pass a decomposing building whose peeling skin reveals adobe bricks. Here's T-ana's Restaurant on the right, pink and brave, and always Mount Blanca dead ahead.

There are about seven miles to cover between the stronghold of the *penitentes* in San Francisco and the meeting hall of the Jehovah's Witnesses in San Pedro. Shonnie grew up between these two buildings. Like bookends, these two slim structures, the *morada* and the Kingdom Hall, enclosed Shonnie's spiritual life, just as her DNA and her suffering body (genotype and phenotype) bracketed the course of her mortal life. Science makes pretty good descriptions of genes and bodies, the biological bookends, when they are looked at in isolation, but to cover all of the contingencies lying between the DNA and a particular outcome in the body—contingencies as crosshatched as Culebra's canals—well, that's another matter altogether.

Modest on the outside but rebellious on the inside, the *morada* and the Kingdom Hall are like Shonnie herself when, to her doctors' surprise, the young woman rejected conventional medical treatment. She was well aware of the risk to her life. What goes on inside a person making that choice? Years before, crying and holding her mother's hand, Shonnie had walked out of the *morada* and into the cloister of the Jehovah's Witnesses. The historical and psychological distance between those two buildings can bring you close to understanding her identity, after which you have to let her free will take over. Even with so decisive a gene as 185delAG, it wasn't predestined that Shonnie would die.

Two events occurred in the mid-1840s that would lead to the *morada* and the Kingdom Hall, respectively. The first was the Mexican-American War. It transformed Hispanos from the masters of their Spanish Catholic realm into a racial and religious minority of the United States. As a consequence, the fraternity of *penitentes* became defensive and the *moradas* of New Mexico became fiercely clandestine sanctuaries of Catholic ritual. The second event was the Great Disappointment of 1844, which was a national event, or rather a nonevent. Evangelical Protestants were convinced that Jesus would return to earth on October 22 of that year, amid rising tribulation. When the Second Coming or Advent failed to take place, the Adventist movement in America splintered. Some of its branches petered out while others increased in energy, until the Jehovah's Witnesses sect emerged in the twentieth century.

The Mexican-American War (1846–48) was basically a landgrab by the stronger country, an exercise of Manifest Destiny. In the early 1800s, the Americans had arrived at the east front of the Sangre de Cristo range, first mountain men seeking furs, followed by traders who laid the Santa Fe Trail. In 1806, the first Anglo to penetrate the San Luis Valley, Zebulon Pike, he of Pike's Peak fame, had been arrested and jailed by the Spaniards. But the Mexican government was accommodating to the foreigners, too accommodating, it would seem. The trigger for war was a dispute over Texas, which doesn't bear recounting here.

As is useful when conducting a war of expansion, the Americans felt morally and racially superior to their enemy. Even before the conflict, the Anglo traders on the Santa Fe Trail were dismayed and sometimes disgusted by the people they encountered. Bad biology appeared to have engendered bad behavior. The Hispanos, one trader wrote, were a "mongrel race . . . miserable in condition . . . despicable in morals." Santa Fe, their squalid capital, was full of beggars and idlers. Even the nobility was dirty and didn't seem to care. Hispano women merited kinder descriptions than the men, but their low virtue was hazardous in another respect. An American in 1850 observed: "Though smoking is repugnant to many ladies, it certainly does enhance the charms of the Mexican senoritas, who, with neatly rolled up shucks [corn-husk cigarettes] between coral lips, perpetrate winning smiles, their magically brilliant eyes the meanwhile searching one's very soul. How dulcet-toned are their voices, which, siren-like, irresistibly draw the willing victim within the giddy vortex of dissipation!"

The blending of races, the spectrum of skin color seething on the streets, excited both horror and fascination on the part of the Americans. W. H. H. Davis, the U.S. attorney for the new territory, noted: "The intermixture between the peasantry and the native tribes of Indians is yet carried on, and there is no present hope of the people improving in color. The system of Indian slavery which exists in the country conduces to this state of things . . . and thus a new stream of dark blood is constantly added to the current." Davis reported that he met light-skinned Castilians, "as light and pure as the sons and daughters of the Anglo-Saxon race." But when the "upper crust" of Taos society was pointed out to him at a dance, he gibed that they were "well baked for upper crust, as a large majority of them were done very brown."

New Mexico became a living laboratory for working out nineteenth-century ideas about the human animal. Anthropologists believed that Europeans and Native Americans represented two of mankind's aboriginal types, the latter having degenerated further from the ideal than the former. When two races so obviously discrepant in physical and mental capacity were merged, their offspring would have to be unstable. According to *Types of Mankind,* an 1854 treatise on race by Dr. J. C. Nott and George R. Gliddon, "Mexican soldiers present the most unequal characters that can be met with anywhere in the world. If some are brave, others are quite the reverse—possessing the basest and most barbarous qualities. This, doubtless, is a result, in part, of the crossings of the races." Also, the Mexican skulls were said to be measurably smaller, a sign of inferiority. Wherever an intermediate race arose, the characteristics of the lower race usually dragged down those of the superior.

Take the example of the mule, a familiar animal in New Mexico, a cross between a male donkey and a female horse. Mules were strong but sterile, which is what you would expect of an intermediate race. The mule's sterility was backhanded proof of the integrity of the parental species, according to scientists of the day. The secondary race of people in New Mexico did not suffer from sterility, far from it, but there were indications that the hybrids were unhealthy. The *genízaro* Indians and mixed breeds at the bottom of the social ladder—the darkest people— seemed to be in the worst shape. The issue for race scientists was not nature versus nurture but the different natures of race, as if culture had nothing to do with a fallen state.

When Anglos added their blood to New Mexico's mixture, the result was not any nicer. J. Ross Browne, in his *Adventures in the Apache Country: A Tour Through Arizona and Sonora* (1869), commented: "The worst of the whole combination of races is that which has the infusion of rascality in it from American sources. Mexican, Indian, and American blood concentrated in one individual makes the very finest specimen of a murderer, thief, or gambler ever seen on the face of the earth. Nothing in human form so utterly depraved can be found elsewhere." Ironically, in accounts of this period, the Pueblo Indians come off as superior to the Mexicans and other half-breeds, since their blood was assumed to be pure (after centuries of cohabitation it could not have been).

So the Americans did to the Hispanos what the Spaniards had done to the Indians—not to equate forms of oppression but to suggest that the Hispanos were put on the receiving end of history and experienced the sting of second-class citizenship. The *casta* system of Old New Mexico surely was racist, yet all members of the society bought into the hierarchy. Along comes a powerful foreign culture holding up its own mirror of racial ranking, and this mirror reveals ridiculous cracks in the Hispano class system. The *españoles* at the top, their threadbare honor mocked, made cringing shows of compliance to the new order, so the Americans in Santa Fe reported. The *genízaros* who were still at the bottom, Marianne Medina's ancestors among them, retreated to the *moradas*.

As Anglos saw it, the biological degeneration of New Mexicans was paralleled by a degeneration in religious practice. In *Death Comes for the Archbishop*, Willa Cather's novel about the period, a character observes, "The old mission churches are in ruins. The few priests are without guidance or discipline. They are lax in religious observance, and some of them live in open concubinage." Another character complains that the people are "full of devotion and faith, and it has nothing to feed upon but the most mistaken superstitions. They remember their prayers all wrong. They cannot read, and since there is no one to instruct them, how can they get it right?" U.S. Attorney Davis, a firsthand observer, was more blunt: "They have an abiding faith in saints and images, and with the mass of the inhabitants their worship appears no more than a blind adoration of these insensible objects." Davis had to inform a local magistrate that citizens no longer could be prosecuted for witchcraft.

But the contrary process, regeneration, might be effected in Hispanos as long as their blood held steady, i.e., by avoiding further miscegenation. Thus Davis concluded his 1857 book about New Mexico with the hope that this benighted people could be raised up by education and by exposure to the American system of justice and democratic principles. Independence, individualism, egalitarianism—all these traits might be taught to them. However patronizing, Davis held the progressive view, allowing a role for nurture alongside nature in human affairs.

Cather's 1927 novel about New Mexico was based on the true-life adventures of a cleric named Jean Baptiste Lamy. A French-born Jesuit, Lamy had been sent by Catholic authorities to reform the Church in Santa Fe. The primitive, hybrid faith created by the Franciscans and Indians would have to go. Lamy arrived in 1851 and got to work, and perhaps it was not coincidental that the San Luis Valley, 130 miles to the north, was settled at this time. As Cather noted, "In lonely sombre villages in the mountains the church decorations were more sombre, the martyrdoms bloodier, the grief of the Virgin more agonized, the figure of Death more terrifying."

Although long claimed by Spain and Mexico, the San Luis Valley had always been too dangerous for permanent settlement. It was, instead, a place for hunting, grazing, and trading with the Indians. The Valley was not safely inhabited until the U.S. Army neutralized the Utes and Jicarilla Apaches.

Just before the Mexican-American War, the Mexican government had made large grants of land in the southern part of the Valley to big developers. These were wealthy individuals who had pledged to recruit settlers and colonize the frontier. One such property was the Sangre de Cristo Grant, a million acres of mountain and prairie held by Charles (Carlos) Beaubien. The United States, on taking over the territory, agreed to honor the Mexican grants and prodded the owners to go ahead with their plans.

In 1851, Hispano *pobladores* from Taos moved north of Costilla

Creek and went over the divide into the drainage of Culebra Creek. (The 185delAG mutation went with them.) They scratched out their village *plazas* near the creek banks and dug *acequias* to catch the water meandering from *la sierra*. Their patron, Beaubien, had staked the settlers to livestock and supplies; they were promised they would earn deeds to their lots eventually. They also were given unfettered rights to hunt game and cut timber on the mountain. The terms of the development must have been attractive, because within ten years seventeen hundred people were living on the Sangre de Cristo Grant.

Meanwhile, gold and silver strikes in the mountains around San Luis Valley brought an influx of Anglos from the East. They were miners, loggers, sheep and cattle ranchers, railroad builders, farmers—the restless resource-extraction team of the American West—but they concentrated their activities in the northern part of the Valley, not affecting Culebra or the Hispanos at first. Alamosa was founded and became a railway hub. Colorado Territory was carved out of the New Mexico and Utah Territories, its southern border hewing to the thirty-seventh parallel. The new border passed directly through the Hispano *plazas* of Costilla, New Mexico, and relegated the Culebra settlements to Colorado.

Each village in Culebra had a holy guardian. La Plaza de San Francisco, established on the eponymous creek in 1853, took Saint Francis for its protector. The settlers built a low, flat-roofed, mud-and-wicker church, and soon afterward they built a *morada*. The original church of San Francisco is long gone, replaced by the spindly white structure described earlier. The first *morada* is gone too, but its replacement still stands and is older than the present church. Every church in the remote villages in the late nineteenth century had a *morada* attached to it. The *morada* watched over the church and its people, not in plain sight but partly cloaked on the outskirts of the village.

Morada is, of course, synonymous with the *penitentes*. Who were these men? No feature of New Mexico history is more controversial than *los hermanos penitentes*. Their formal name, *La Fraternidad Piadosa de Nuestro Padre Jesus Nazareno*, means The Pious Fraternity of Our Father Jesus the Nazarene. In one sense they were no different from nineteenth century contemporaries like the Freemasons, Odd Fellows, and Knights of Columbus. They were an all-male club that did charitable work in the community while keeping their internal operations secret. They settled disputes, looked after the sick, organized funerals, and raised money for the poor. But more than that, the *penitentes* sustained the religious life of the villages in the absence of priests. Their most prominent duty—which they never yielded to the Catholic reformers in Santa Fe—was to lead the annual Passion procession.

On Good Friday in Holy Week, two groups of *penitentes*, called the Brothers of Blood and the Brothers of Light, would emerge from the shuttered *morada* of San Francisco. They would head down the road toward the church. The Brothers of Blood, who were the young initiates of the society, bare-chested and wearing black hoods, walked at the front of the procession. Several dragged crosses . . . one pulled a cart with the mannequin of Lady Death . . . the rest scourged their backs with rhythmic lashes, first over one shoulder, then the other, a score of whips cracking as one. The whips (*disciplinas*) were made from boiled yucca leaves, an Indian technique that made the leaves flexible but kept the edges sharp. Some of the young *hermanos* might have cholla cactus strapped to their backs, the spines digging in with each step. Next came the Brothers of Light, wearing white, the older men and officers of the society. Holding torches, they sang the *albados*, the slogging hymns of regret and penance. The whole village attended the pageant, falling into line as the brothers passed and then turning back from the church to the *calvario*, the hill above the *morada*, where a mock crucifixion would take place.

As night crept up the mountains, the Passion ritual moved indoors for the climactic ceremony of *tinieblas* (darkness). It was the only time the public was invited into the *morada*. The elongated building had three rooms: on the left a chapel with a wooden altar; a central space where the brothers had their meals and socialized; and on the right a private room for rituals, the *kiva*, as it were, with a rounded fireplace in the corner. On this night the private room was locked; people took seats in the chapel and in the center room. A triangular-shaped candelabra provided the only light—six candles on each ascending leg of the triangle, to represent the twelve apostles, and two additional candles on the base, standing for the Holy Mother and Mary Magdalene. One by one, the candles were snuffed out to mark Christ's drawn-out agony. The sound of the lashes and the crying of the *albados* were much louder when heard indoors.

Los hermanos penitentes did not spring out of nowhere. Europe throughout the Middle Ages had roving societies of flagellants. A Portuguese traveler in Valladolid, Spain, in 1605 described a pageant of two thousand flagellants bleeding copiously as they paraded before the king. To whip yourself as Jesus was whipped represented an extreme trial of the flesh, the truest imitation of the Savior. The penitential societies were held at arm's length by the Church, however. Catholic leaders recognized the risk of overdoing the pain for the holy pleasure of it. Saint John of the Cross called such practices spiritual gluttony. Saint Francis and Saint Teresa of Avila forbade their followers to mortify their bodies excessively. Occasional scourging was fine, but drawing blood gave asceticism a bad name. Fasting, mixing your food with ashes to dull the taste, exposing the body to cold and heat, feeling the rough wool habit or hair shirt (*cilicio*) against your skin—these ought to be sufficient daily reminders of the body's sinfulness.

Franciscan priests introduced corporal penance to New Mexico. The members of Juan de Oñate's discovery party, marching north, are said to have flagellated themselves on Holy Thursday in 1598. But the organized lay brotherhoods didn't appear on the landscape until the 1830s, after the withdrawal of the Spanish *padres* created a clerical vacuum in the villages. Unschooled and crude though they were, the *penitentes* performed a social service. They could be regarded as a normal extension of Hispano culture until the Americans came, and then, in the mirror held up by the *gringos*, they were turned into freaks. Prohibited by the bishop in Santa Fe from marching with their crosses or lashing themselves in public, the *penitentes* went underground. In private political protest, they seem to have stepped up their penances and also to have toughened the initiation rites for new members. It was rumored that they held secret crucifixions in the hills using real nails, not ropes.

By the early twentieth century the New Mexico *penitentes* were part of the pulp fiction of the American West. Tourists came hoping to get a glimpse of their sensational rites. A 1936 B movie, *The Lash of the Penitentes*, showed a naked woman being whipped in a *morada*. Though it purported to be a documentary, the film exploited both racial and religious prejudices against Hispanos. "Wake up, America!" the scandalized narrator intoned. "Here in our own country we can see the very heart of Africa pounding against the ribs of the Rockies!"

But after World War II, the melting pot of assimilation weakened the *penitentes'* appeal. Hispano boys who had joined the army were embarrassed when they were asked to explain the scars on their backs. The Church made amends with the *moradas*, ending the sanctions, and women were allowed in as auxiliary members. Flagellation—the word itself went out of use, let alone the practice. From about three hundred brothers at its peak, the roll of the San Francisco *morada* shrank steadily, along with a decline in churchgoing everywhere. Thus tolerance, on the one hand, and an easing of devotion, on the other, normalized the *penitentes* and made them unexceptional figures in Culebra once again. Tepid Catholics such as Joe U. Medina became members, and his son Joseph, equally listless in his faith. As Maurice Fishberg might have said, the *penitentes* prospered best under the iron rule of isolation.

In 2000, the San Francisco *morada*, in urgent need of repair, was added to the Colorado Register of Historic Properties. The State Historical Fund awarded a grant for interior and exterior restoration. An architect was hired, a new roof was put on, and the crumbling stucco was replaced by a mud-based plaster, such as would have been used in the nineteenth century. The restored building, painted brown, stands out from the hillside more than it ever used to. Things have really changed when you can ask a member of the brotherhood to show you into the *morada* and he cheerfully agrees to do it. With one or two exceptions— serious older guys who keep their mouths shut about their practices— the *penitentes* have been consigned to the specimen jars of history.

———

Shonnie and Iona squeezed themselves next to Marianne in the crude chapel. When the last candle went out and the room was black, a voice said, All living and dead come forth to join us for the love of God. Men stamped their feet, pounded the walls, and dragged chains across the floor, invoking the earthquake and the chaos that broke out in Jerusalem when Jesus gave up the ghost. A demon materialized, spinning his *matraca* in the beam of a flashlight. That noise, a grinding thing—it was horrible, said Marianne. My girls were screaming.

When it was over, someone lit the candles again. Filing out of the *morada*, people smiled as they do after a good performance, and readied their spirits for Easter, but Marianne and her kids had had enough of Catholic ritual. Without knowing it yet, they were headed for the Kingdom Hall.

Protestant missionaries first came to New Mexico during the last quarter of the nineteenth century. Presbyterian, Methodist, and Baptist, they went door-to-door distributing Spanish Bibles to families who had never owned one, dodging gunfire from Catholics on a couple of occasions. The Presbyterians were the most active in seeking Hispano converts, and they counted some success even among the *penitentes*. Anglo missionaries in the Southwest spoke scornfully of Catholicism and its gilt-edged twisting of scripture. Bishop Lamy in Santa Fe countered that the Catholic Church had never barred its people from owning Bibles, but that indeed it was forbidden to interpret the Old and New Testaments without proper guidance.

Protestantism made small, tenacious inroads in the Valley and along Culebra Creek. In the 1890s, a few families in the hamlet of San Pablo flipped to Presbyterianism and were ostracized by the Catholic families in San Pedro, on the other side of the creek. The cemetery shared by San Pedro and San Pablo was fenced down the middle in order to keep the deceased of the two churches apart. The same split occurred at the cemetery in Los Fuertes, a few miles down the road. Other Protestant groups came and roosted on the dry prairie west of Culebra. The Mormons established farming towns at Manassa, Sanford, and Eastdale. A colony of Adventists built the town of Jarosa nearby.

The Protestants were by nature schismatic. Where Saint Teresa and the Catholic reformers took their disagreements with the Church inside themselves, retreating to the personal *moradas* of their faith, Protestant reformers moved in the opposite direction, wrenching free of the gravity of Rome and subdividing again and again. Paradoxically, each far-out faction claimed to have reverted to the elemental core of Christianity. No high priests were necessary for understanding the Bible, no inter-mediating saints—on that all Protestants were united—but the Bible had so many facets that every side could find verses supporting its own particular doctrine.

On the question of the end times or last days, Protestants could be sorted between progressives and conservatives. The Presbyterian missionaries who were saving Hispano souls in San Luis Valley believed in human progress in anticipation of the Millennium. They believed that the world could and should be improved before the predicted Second Coming of Christ. Indeed, America was the very place on earth, God's chosen place, for making society right. This optimistic vision of human regeneration did not worry very much about time or deadlines. A thousand years might see the job done, after which Christ would descend as icing on the cake.

The Adventists and other fundamentalist Protestants were pessimistic about the human species. They thought that the human races were immutable and that striving for social progress was pointless. When Jesus came, which would be soon, there would be Armageddon; the battle and fire would take care of the improvement that was needed. Believers of this stripe had gotten over their disappointment that Christ had not returned in 1844. After consulting the Bible and finding errors in the previous calculations, the Adventists pinned their hopes on 1874. Let down a second time, they made a technical fix: Christ *had* returned, only invisibly. Now they needed to know, When would Armageddon begin?

A self-taught preacher named Charles Taze Russell split from the

Adventists. He and his adherents called themselves Bible Students. Parsing the Books of Daniel, Revelation, and others, Russell figured a date of 1914 for the end of the present world and the inauguration of the next. World War I made a good start on his prediction, but then things on earth stalled. Russell died, part of his following peeled off, and the remaining Bible Students reorganized themselves as Jehovah's Witnesses. The Witnesses assigned Christ's invisible return to 1914. Avoiding further forecasts, they have been on hold for the main events of Armageddon ever since. Certainly the twentieth century provided plenty of rumblings.

Jehovah's Witnesses are known for knocking on your door unexpectedly, and the fact that they are dressed formally and outnumber you two to one puts you on notice immediately, as if a pair of IRS agents had announced a surprise spiritual audit. Scrubbed and self-assured, they seek to engage you, and if necessary debate you, on urgent matters of the Bible. They don't get mad if you demur but will back off politely with the intention to try you again on another day. Well, a look might pass across one of their faces, suggesting, Don't say I didn't warn you. As the Gospel advises, they shake the dust off their feet and walk on to the next house.

Though sometimes lumped with them, Jehovah's Witnesses disapprove of charismatic or Pentecostal Protestants because of their sweaty emotion and born-again spectacles, their prostrate crying to Jesus for forgiveness and succor. No, for a Jehovah's Witness the correct course of religious experience is to coolly, closely study the Bible and then live by its lessons. They study with the aid of their monthly magazine, *The Watchtower*. The topics of the articles in *The Watchtower*—Jesus, global warming, the Internet, personal problem-solving, the ancient Israelites, Satan, health, "Marriage and Parenthood in This Time of the End," maintaining one's integrity, the fate of the dead after they die, the current Witness ministry in South Korea, etc.—have a timeless, almost dreamy variety to them. A quaint jumble is made of the past, present, and apocalyptic future. In the illustrations, King David, Paul the Apostle, and other well-muscled, kindly figures of antiquity rub shoulders with everyday moderns like you and me and also with denizens of the coming Paradise, all looking pretty much the same except for their dress.

The information and advice given in *The Watchtower* are meant to be so comprehensive that you do not need to read anything else, other than the Bible itself, of which the Witnesses have made their own translation. Adding to the sense of anachronism, their Bible is divided between the Hebrew-Aramaic Scriptures and the Christian Greek Scriptures rather than the Old and New Testaments. *The Watchtower* treats the Bible not as a Judeo-Christian narrative but as a grab bag of quotations retrodictively confirming the Witnesses' views of the universe. Learning at the meetings isn't about sifting and weighing ideas but rather the memorization of talking points, led by an elder.

The whirling souls of the Jehovah's Witnesses arrived in Culebra during the mid-1970s. Parking on the side of the road and walking systematically from house to house, they stuck out like sore thumbs in the community. Joseph Medina didn't like to see them; he chased them away from his door. He was downright mean to them, recalled Marianne. The young couple had returned to Culebra following Joseph's stint of playing rock music in Denver. With their two little girls, the Medinas expected to resume their lives as Catholics. It's a tradition, said Joseph. I figured, I was raised in it, my kids will be raised in it. I was content with being a Catholic.

Marianne didn't care for the Witnesses because most of them were white. As a rule she wouldn't permit Anglos inside her house. On the other hand, she considered Catholicism a religion that had been forced on the Native Americans. We lived the name but didn't believe strongly, she said of her Catholic Indian relatives. We went along because we didn't want to be put down. Since they had been back in Culebra, Joseph's father had been after the family to show up at the San Francisco *morada* on Good Friday. To keep the peace, we went, said Marianne, grimacing at the memory. That was the only time.

About then, a Hispano family that had moved to another state and converted to the Witnesses decided to return home. The family befriended Chavela Medina, Marianne's sister-in-law. Chavela started studying the Bible with them. And then *I* started studying, said Marianne. I knew the *penitentes* weren't right, and I knew the Catholics weren't right. But I was not easily convinced.

Marianne fact-checked the Witnesses' Bible against the standard versions she found in the library. She was happy to discover that God indeed had a name, Yahweh aka Jehovah. She was able to show her skeptical husband that the passages highlighted in *The Watchtower* were also in the Catholics' Bible—yet the priests ignored these verses. Joseph began to investigate the Bible on his own accord. I found that hellfire isn't in it, he said. I found that the saints aren't in it. Here we are calling priests Father—that's not in it either. Joseph went to the priest in San Luis and got him to admit that God's name was Jehovah.

One of the knocks on the *penitentes* in the nineteenth century was that New Mexicans earned the right to do whatever they pleased for eleven months of the year in exchange for whipping themselves during Lent. When Joseph examined his own behavior, he saw something similar. For Catholics it's OK to do bad things and then confess to priests, he said. I was trying to get away with what I could. I didn't care. It's OK to cheat a little, lie a little. It's all right to take things if you have to. It's all right because God will forgive you. You can get drunk once in a while. . . .

On October 18, 1980, Joseph and Marianne Medina and a dozen other converts from around the state were baptized in a high school swimming pool in Pueblo, Colorado. The Jehovah's Witnesses had rented the pool for the afternoon. Shonnie, who was ten, and Iona, seven, peeked through the slats of the fence and giggled at their father's skinny white calves. The two girls had never seen Joseph's bare legs before. It was a modest, earnest ceremony, the men as well as the women wearing T-shirts over their bathing suits as they were dunked in the water.

Before long, Joseph's sisters Chavela, Lupita, Louisa, and Wanda also were baptized, along with their mother, Dorothy, and two of the brothers. One of us learned and we tried to teach the others, said Lupita.

When I saw it was true, I thought it would help all of us. I could see the difference right away, she continued. You read it and you applied it. The Witnesses are family-oriented. You feel part of a family. Lupita contrasted the warmth of belonging to the Witnesses with the hypocrisy of her childhood Catholic worship. I remember what it was like, sitting in church, she said. The empty words, the requests for money. We can't afford shoes, how can we give to the church? In the pews you hear, Look at who's here! And, Look at what they're wearing! It gives me chills to think of it.

But Culebra's reaction to the Medinas' conversion was hostile, a reprise of the reaction to the Anglo missionaries. A buddy of Joseph and Marianne's, a former bandmate who lived close by, blasted music at their house, taunting them with songs the band used to play. When Iona's horse died, apparently poisoned, and Joseph's dog was shot, Marianne was certain it was because of their conversion. The priest took it badly too, and publicly washed his hands of the family. It didn't help that Joseph and Marianne turned around and attempted to convert their neighbors. Going door-to-door with copies of *The Watchtower*, they half-exulted as the doors were slammed in their faces. We were zealots when we worked La Valley, Marianne admitted. We needed to calm down.

By the mid-1980s, there were enough Jehovah's Witnesses in and around Culebra to warrant the construction of a Kingdom Hall. A site was found in San Pedro, near San Luis. The men of the congregation poured a concrete slab foundation, and then, with the help of dozens of Witnesses summoned from outside the Valley, they built the Kingdom Hall all at once, as in a barn raising. It's done fast to show unity, Marianne explained. Everyone has to help. The frame was up in twenty-four hours, the crews working around the clock in spite of snow. Teams of volunteers were organized for drywall, painting, carpet-laying, cooking, errands, etc. Shonnie and Iona helped by passing out snacks. The Medina household was full of people. The visitors slept in campers or RVs parked in the local Witnesses' yards. When the priest came by the site to look, he shook his head ruefully, according to Marianne, saying, Our members should be this united.

At intervals as the Kingdom Hall materialized, someone would snap a photo of a boy standing in front of the project. He held the large wooden hands of a clock, as if to warn people that the end was near. It was, in a benign sense, because the building was completed in just three days, inside and out. The stop-action timekeeper was Michael, Shonnie's future husband. The two would marry here seven years later. The first wedding was that of Shonnie's aunt Wanda to Bill Kramer, in the summer of 1985.

The San Luis Kingdom Hall of Jehovah's Witnesses sits back from the road, tucked behind a gas station and liquor store, seven miles from the *morada*. You might pass it a hundred times without seeing it. The Kingdom Hall does the *morada* one better on the question of windows. Where the *morada*'s are heavily shuttered, never opened, suggesting somber and mysterious rites, the whitewashed Kingdom Hall has no windows at all, owing to concerns about vandalism. Yet inside there is plenty of illumination, and cheerful pink walls, and rows of comfortable seating. In place of an altar or cross is a soaring mural of alpine scenery, set off by vases of silk flowers. There are no images of Christ or holy personages. The Witnesses shun religious icons because the Bible bans idolatry. Even the Cross they deny because they maintain that Jesus was sacrificed on a pole or stake.

Christianity historically has enmeshed itself with state power, but the Jehovah's Witnesses operate a theocracy that ignores the powers of state. Having withdrawn from the world as far as they are able to legally—they do not vote, or hold military or political office, or acknowledge national or religious holidays—they have made a facsimile of the world inside the Kingdom Hall, where, cut off from daylight's cues, you have the slightly dizzy feeling that the new system has come down to earth already. If Armageddon were raging outside, here in their hermetic company you'd be safe.

Marianne once remarked that people who are drawn to the Witnesses tend to have bad pasts. She would count as elements of a bad past the racial slurs she suffered growing up in Alamosa, or the beatings Joseph and his siblings received from their father when he was drinking. A bad past is a past that persists and grates. Witnesses try not to get emotional about their faith because strong emotions are scary—passionate feelings rip the scabs off cuts that should be left alone. Wanda's husband, Bill, spoke of himself as physically and sexually abused. I had a sore spot on my heart, he said, and he meant an actual sore from the abuse. For people whose hearts are bruised, the Kingdom Hall is a refuge and the end times may not come soon enough. In the new system, they believe, their hearts will be healed and restored to them eternally.

But Shonnie, the radiant young adherent, who was tutored in the faith and was baptized at seventeen, belongs in a different category, doesn't she? Life may have wounded her parents, but they repaired themselves and protected Shonnie from the same harms. True, they could not prevent her being molested by her uncle, a sorry little man she later forgave, nor block her inklings about dying young. Still, Shonnie had a golden childhood. Her bad past, if she had one, was symbolized by the *morada*, with its showy and shallow secrets, its dalliance with pain and death. It was simple for the child to let go of Catholicism. Her family had hardly any Catholicism left. If culture provides the atmosphere for identity, Shonnie assumed a Witness identity as naturally as she breathed.

What's more, inside the Kingdom Hall the racial history of New Mexico no longer rankled. Shonnie was postracial, to use a term not then in use. A defining characteristic of a religious sect is that the differences among the members are outweighed by the group's differences with outsiders. By the time the Medinas joined, the Jehovah's Witnesses had evolved into a tolerant, multihued organization, claiming one million members in the United States and seven million worldwide. There is the same type of personality of Witnesses the world over, Iona noted. She and her husband were close friends with Witnesses in Australia. Everybody's equal, Wanda said, citing the mixed marriages in the congregation. Marianne Medina credited the Witnesses for softening her hostility toward whites.

Besides holding a Sunday meeting at the Kingdom Hall for Bible study, the local Witnesses meet during the week for what they call Theocratic Ministry School. This is to go over techniques for interacting with the public and improving their door-to-door preaching. Tonight, a Tuesday, Joseph Medina was scheduled to offer his insights. He was one of the three male elders (there are no clergy or female elders) of the little congregation, about two dozen Witnesses in all. He wore a charcoal suit with wide lapels and full-cut trousers. His wife and daughter were among the dozen or so in the audience. Marianne, demurely wearing glasses, had her hair up, and Iona was regal in a long black skirt and white blouse. Bill and Wanda were there too, Bill fidgeting and fingering his Bible.

Joseph went up to the stage. His head didn't reach far above the lectern. Use your gift, was the theme of his soft-spoken talk. We want to reach people's hearts. Listen and respond. We glorify God as we help others. It was hard to imagine Joseph as a zealot for Jehovah without Marianne standing beside him. The other two elders, named Brother Rivera and Brother Alonzo, had a more polished manner. Passing the microphone between them, they thanked Brother Medina for his contribution and sought feedback from the others on what they had learned about going door-to-door.

Shonnie had been a born evangelist. She was very confident when ringing a doorbell because she was used to making a good impression on strangers. Hair, makeup, clothes, the works. Plus she had a light touch. Before she was married, Shonnie was paired for her preaching service with a woman named Claudia, who was a rather self-righteous Witness. The two would pull into a neighborhood, park, and, while freshening her lipstick, Shonnie would sing along with the radio, doo-wop music especially, the kind of songs her father liked. One day the older woman tried to bring Shonnie down a peg, saying, Should we be listening to that? Shonnie just laughed and said, Oh, Claudia, live a little!

Witnesses who were in good standing kept track of their hours of service. If their timesheets showed they preached an average of ninety hours per month, they were awarded the status of Regular Pioneer. For seventy hours of service they would get to be an Auxiliary Pioneer. Shonnie made Regular Pioneer and hungered to do more. Alamosa, where she moved after marrying, was more fertile ground than Culebra, because it was bigger and had a less rooted population. On the weekends, after putting some meat and rice into an all-day cooker, Shonnie and Michael would go forth with their copies of *The Watchtower* and the golden answer book, *What Does the Bible* Really *Teach*?

Everyone agrees that service was the highest priority in Shonnie's life, after being a good wife. Even when she was sick, she'd go out, said Marianne. Even in the hospital or the clinic she had leaflets and books to give to the nurses and to the security guards. Her mother doesn't remember anyone getting irritated with Shonnie when she preached, because she had charisma, said Marianne. Shonnie seems to have been a woman who was fully conscious of her charm and beauty without allowing it to go to her head. She acted as though she could shine her light on the people she met, and they would become beautiful too, but not by her reflected glow—they'd awaken and sparkle spontaneously. Use your gift, her father had urged.

She'd take the energy and shine it out, her uncle Bill said. She was a person of innocence and wonder. She had a balanced way of being. She made life look easy, but she was humble too. Bill remembered how Shonnie always saw the good in everyone. She helped him particularly after he married into the family. Sometimes I felt demeaned, he said. Uncle Bill, I spoke up for you, she'd say. Things will work out. She was only fifteen. And if Bill was talking too much, getting agitated, Shonnie would give him a hug and he'd calm down.

Admittedly, Shonnie is not so much a character in this narrative as an avatar, like Quixote's Dulcinea. The words and deeds of others in her family could be tested by observation, but Shonnie's character had slipped behind the veil of memory, refulgent and half-beatified. As her mother once put it, If there was a black spot or black mark on a white sheet of paper, she'd say, Look at the white, all the white around it. That was Shonnie.

Since the time when Shonnie was active in the Jehovah's Witnesses, the benchmarks for service have been relaxed, so that just seventy hours a month of preaching makes you a Regular Pioneer. Iona was a Regular, and Marianne an Auxiliary, and occasionally they worked the small towns of the Valley together. Since they had covered the same territory many times, they knew which houses to avoid and which were moderately receptive to their message.

Tonight, as a demonstration for the service meeting, Marianne and Iona acted out the roles of a model Pioneer and a typical householder, the person answering the door. Mother and daughter went up to the stage. The scenario was this: The householder, played by Marianne, was a young person who needed help with her science homework. She had to write a paper on astronomy. The Witness, who was Iona, steered the conversation toward the Bible and the remarkable harmony, said Iona, between the Bible and modern science.

Iona, as she consulted her text, did most of the talking, while Marianne's part was mainly to nod. The issue before the student was the shape of the earth, whether it was flat or round. The Witness commentary maintained that, although philosophers and scientists had long declared the earth round, no one believed it until the late twentieth century, when astronauts went to the moon and beheld the sphere of Earth with their own eyes. This reading of history greatly exaggerates the influence of the flat-earth lobby, a renegade minority for more than two thousand years. But Iona's objective was to contrast the scientists' uncertainty with the sure statement in the Book of Isaiah (40:22): "There is One who is dwelling above the circle of the earth, the dwellers in which are as grasshoppers." In referring to the circle of the earth, the prophet knew that the earth was round—and so you see that the Bible had it right all along. Bill Kramer was not the only person fidgeting in the Kingdom Hall that night.

Jehovah's Witnesses do not promote higher education. The time spent on campus might be wasted as Earth hastens to its upheaval, but more immediately, the university exposes a young person to selfish competition and materialism, to say nothing of drugs, alcohol, and the questioning of religious authority. Better for the student to take practical, vocational courses and let the Bible supply her higher education. Shonnie did well enough in high school to go to college and could have received financial aid, according to Marianne. But even before graduating she seems to have withdrawn from student affairs, eschewing a class ring and team sports and group outings. Shonnie liked riding, singing, reading, and of course her preaching, all of which she could do apart from school.

After graduation Shonnie took a few courses at home by mail, looking to become an accountant perhaps. She also considered becoming an aerobics instructor. When she got a job at Maurices, a women's fashion store in Alamosa, she discovered her secular calling. The skills she used in her Witness ministry proved very effective in selling clothing and makeup. Shonnie won sales awards from Maurices. She was so successful that the store managers had to limit her hours because other salesgirls complained that they were losing customers to her. Shonnie later became a direct retailer of Mary Kay cosmetics and won awards from that company also, not quite earning the fabled Mary Kay pink Cadillac for her sales, but close. Shonnie used to model Mary Kay products at store openings, fairs, and other events. Women thought, If I buy from her, I'll look like her, said Marianne.

The boys who pursued her bumped into an invisible shield, her chaste gloss, because if they weren't Witnesses, forget about dating Shonnie Medina. Dating—there was no dating or kissing without the intention or goal of marrying. It's not a game, said Marianne. Shonnie did not date until she met Michael. Their courtship wasn't as scrutinized as, for example, an Orthodox Jewish *shidduch*, which involves a matchmaker and squads of relatives, but Shonnie and Michael always had chaperones, and when they went to the movies, they had to travel to the theater in separate cars. When she was a married woman in her twenties, Shonnie would take teenage girls under her wing and counsel them to act appropriately toward boys. They paid attention because the girls could see how boys looked at Shonnie. A friend of hers named Jackie recalls, I was still in the world when I started studying with Shonnie. Fortunately, I'd never committed immorality. She was firm. She was clear with me.

To compare her circle of Jehovah's Witnesses with fundamentalist Jews is apt. Ignore for the moment 185delAG, the genetic link between the Jewish people and the fated family of Culebra. In their preoccupation with scripture, their strict morals, their endogamous marriages, their wariness of wider society, their avoidance of blood in food preparation, and their lack of interest in the desires of the soul, these two religious minorities are more alike than any in America. Jehovah's Witnesses constantly discuss the Jews, although not as Jews per se but as the original worshippers of Jehovah, steadfast in their faith, refusing to surrender their identity—in short, Jews as proto-Witnesses. Supersessionism, a venerable argument in Christian theology, holds that the Jews botched their covenant with God by failing to recognize Jesus, and that Christians superseded them as the chosen people. The Witnesses think they have earned the mantle of the Israelites. They think the Bible's prophecies, from Genesis to Revelation, point solely to them.

The similarities between the two sects break down when it comes to science, however, for even the most Orthodox Jews respect and encourage science. They do not take the Bible's ban on consuming blood so literally as to refuse blood transfusions, as Witnesses do. They look upon God's creation as a puzzle to be solved, not a holy mystery to be admired from a safe and trembling distance.

The figure of Jesus distracts ultra-Christians from the composition of the world. Intervening between man and nature, Jesus blinds the faithful with his aura. Because of the supernatural exhibitions of Jesus, culminating in his rising from the dead, and because he also promised that he would come again at an unknown time, the strictest Christians keep their eyes on the horizon. If the world will soon pass away, they have less of a need to understand it and more of a need to know what will replace it. Jews expect a Messiah too, but, never having had a taste of him, they don't let their vision of the Messiah interrupt their tasks on earth. Which goes to say, they don't let faith get in the way of reason.

When Shonnie consented to marry Michael, her father took her aside and spoke to her about the importance of headship, another imperative that the Witnesses share with strict Orthodox Jews. It means that the husband is in charge of the couple's decisions, which he will make after consulting and honoring his wife. With good reason, Joseph Medina was concerned that Shonnie would balk at the arrangement: Joe said I wasn't docile, said Marianne with a smile, and Shonnie was headstrong like me.

According to Marianne, there was a joke among the Witnesses that if the husband was the head, the wife was the neck (because one turns the other). That's not funny, Joseph retorted. When the meeting at the Kingdom Hall concluded this evening, Joseph and Marianne stood for the final prayer. Holding hands, they tilted their heads toward each another until their temples touched.

Shonnie and Michael entered married life resolving not to have children. The world was not suitable for children, they felt, and Iona and her husband had the same view. I don't want to have them, Iona said. We both decided, we don't want to raise kids in the system we have now. We'll wait until the new system. Then I'll have as many as we want, in a better world. Marianne, the tough-minded mother, said, Would you put your money into a business you knew was bankrupt? This system is going out of business. There's no future for anybody in this system. It's coming down—that's why we don't get involved in politics.

But wasn't that terribly pessimistic? To conceive of the world so darkly? It's not the *world* that's coming to an end, Marianne explained. It's the system. Remember, there's Paradise. . . . If you have a house with mice, you won't burn it down, your house. You'll get rid of the mice.

The turmoil had already begun, thunder on the horizon from earthquakes and plague. When at last Jesus takes the field against Satan, the polluters, the wicked people of the planet, all will be destroyed, they believe. Witnesses do not sit on their hands in expectation of that day. They work to save others, house by house, neighborhood by neighborhood, but they don't try to scare you. When they are not out proselytizing or in the Kingdom Hall studying, they are as relaxed and fun-loving as any people can be. Celina Gallegos, another friend of Shonnie's, said, People focus on the destruction but there's also the *relief* of Armageddon. All the crime, all the violence, these things which we know like the time of day—it's over. Jehovah stops it.

Shonnie would not know about the glorious circumstances of the end because she was suspended in a state of sleep. She was not dead, said Celina, she was asleep. When she awoke, her body resurrected, her beauty redeemed, she would want to hear all about Armageddon from those who had lived through it. For there will be survivors, Celina insisted. The potential is for any one of her family members to be a survivor, Celina said.

While she lived, Shonnie in her windowless Kingdom Hall did not feel excluded from society but perfectly included, free to pick and choose from the world's transient offerings. She did not dismiss scientific medicine, but she was not convinced she should value it over any other kind of medicine. Her attitude about the doctors, scientists, politicians, and other secular authority figures was that they were standing on thin ice, centered on the wrong things. This opinion about the world was a received opinion. A large part of her identity was a group identity, which she augmented with the stubbornness of her father and the outspokenness of her mother. Like a *penitente*, Shonnie was not afraid to go against the grain, especially where the body and its treatment were concerned.

Finally, it is worth stating that Witnesses do not believe in predestination or in the unceasing torment of the damned. Rather, they think the individual can and must take morality into her own hands, choosing the right way to live, and thereby escape death. So that was Shonnie just before she got sick: a Pioneer, a wife, a model, an unwitting BRCA carrier. Between the *morada* and the Kingdom Hall, forceful and trusting, traveling on her own free will.

Chapter 7

THE DNA AGE

fter Watson and Crick discovered how the DNA molecule works, in 1953, many questions about life could be answered, along with a big metaphysical question. The dualism between mind and body, spirit and anatomy, soul and flesh, could finally be put to rest. DNA, the thinking man's microanatomy, honed by evolution and sensitive to signals from the landscape, operated both the body and the brain without any metaphysical assistance.

Masterminding life from deep within the cell, deoxyribonucleic acid (DNA) performed two tasks, one having to do with form and the other with function. The first task of DNA was to replicate itself and control heredity, thereby preserving the form that enabled the genetic content. The second task was to commission the manufacture of proteins by means of the genetic code. Proteins were the workhorses of the body and were engaged at every level of activity, both physical and mental. Genes and proteins could help explain the problem of the soul. In DNA's model of human nature, a restless soul was but an agitation of the neurotransmitters, which were a class of proteins ordered up by the genes in response to unsettling cues from the environment.

Or maybe a restless soul did not operate just that way genetically; it might be a mendelian trait instead. As Mendel sensed, traits like this are produced by powerful, single genes. Tay-Sachs disease and heritable breast and ovarian cancer (HBOC) are familiar mendelian traits, as are benign attributes such as cleft chins. Genes of this sort put their stamp on people at birth with no prompting from the external environment—they pay no heed to the landscape. The original term in medical genetics for such traits was inborn errors of metabolism. Now, if a restless soul were a mendelian trait, it would probably be a recessive one. The condition would be inherited from two ostensibly imperturbable parents, each of whom carried the tendency without knowing it. In the DNA age, you could imagine a genetic test for trouble of the soul, which might be offered to sufferers or potential carriers or both. However, most people would probably not want such a test unless a treatment for the soul were available also.

Early in the DNA age (to return to actual history), the technology was not good enough to detect genes directly, so medical geneticists worked backward from proteins. Proteins that had gone awry in certain families were indicators that genes must have gone awry. The mendelian genes in question were altered by mutations, and usually the mutations cohabited with copies of genes that worked normally. Remember that genes come in pairs. For carriers of recessive conditions like Tay-Sachs, the healthy copy of the gene prevails, and the carrier is normal, but the disease will strike in the next generation if two carriers mate and their disabled genes come together. In a dominant condition, the malfunctioning gene overrides the healthy copy, in some cases crippling the protein, in other cases leaving the healthy gene open to attack, which is what happens to carriers of BRCA mutations.

Heritable breast and ovarian cancer and its lead agent, BRCA1.185delAG, came to light relatively late in the DNA age because, as with the Hispano cases, the inherited cancers were masked by the more numerous, sporadic types of the same conditions. BRCA and HBOC have since become special concerns for Ashkenazi Jews, but before they knew about BRCA, they had to deal with Tay-Sachs disease, prefiguring the group's response to BRCA disease. Tay-Sachs has been mentioned several times. Although the condition was rare, affecting fewer than one in two thousand Jewish babies, the suffering of the children as their mus-

cles failed was unbearable to families, and the predictability of the inheritance pattern was maddening to the community's doctors. If the carriers
could only recognize the Tay-Sachs genotype, the phenotype would have
no place to hide. Like a dybbuk wrongfully possessing a soul, the disease
could be exposed and perhaps eliminated.

Tay-Sachs disease is caused by the failure of a gene to make a critical
enzyme (an enzyme is a type of protein). In the 1960s, long before the
mutation's location on the DNA was pinpointed, researchers were able to
measure the enzyme, or lack of it, in blood. A Tay-Sachs carrier expressed
about half the normal amount of enzyme, making enough protein to pre-
clude the disease, but an affected child showed no enzyme at all.

Population screening for the enzyme began in the early 1970s. First
in Baltimore and Washington, DC, and then in other cities, Jewish men
and women found out whether or not they were Tay-Sachs carriers. The
turnout was extraordinary. Fired by pride and educated to the need by
their rabbis, doctors, and local health departments, thousands gathered
at synagogues to be tested. Eighteen hundred men and women braved
the rain in Bethesda, Maryland, one Sunday in May 1971. On a single
day in Riverdale, New York, in 1975, Harry Ostrer and a medical-school
classmate drew blood from five hundred people. The experience, which
reminded young Dr. Ostrer of the outpouring of Jewish pride during
the 1967 Yom Kippur War, inspired his decision to specialize in medical
genetics.

Local hospitals responded quickly to the screening programs. The
pregnancies of carrier couples could be monitored by amniocentesis and
terminated if the fetus was affected. Soon the statistics registered a drop
in Tay-Sachs from about sixty cases a year among Ashkenazim to fewer
than ten. And by the turn of the twenty-first century, a million and a half
American Jews had undergone carrier testing. Hundreds of fetuses were
aborted, it is true, but during the same period some twenty-five hundred
children were born to couples in which both husband and wife were carriers. Group screening having accomplished all it could, the testing for
Tay-Sachs takes place today in outpatient clinics, on college campuses,
and via online services. The handful of cases that still occur are most

likely to affect non-Jews—that is, families carrying mutations other than the distinctly Ashkenazi variants of the Tay-Sachs gene.

Fundamentalist Jews were the last branch of Ashkenazim to take advantage of Tay-Sachs screening. The Orthodox and especially the ultraobservant Hasidic Jews considered childbearing an essential duty not to be interfered with. Carrier testing not only opened the door to contraception and abortion but also could destroy a person's prospects of marrying. In the close-knit urban wards of the Hasidim, where men dressed in black suits, women covered their hair, and young people were funneled into marriage via the *shidduch*, the rabbi wielded more authority than the doctor. Medical information was not private. When two families were sizing up a match, just the whiff of a genetic problem could taint everyone on one side. The concern with stigmatization cannot be underestimated, said Harry Ostrer. Once the neighbors hear that Shmuel has been to the geneticist, Harry explained, they assume the worst about his having a genetic condition and might conduct a whisper campaign.

Orthodox rabbis therefore refused to cooperate with Tay-Sachs screening. In the early 1980s, Josef Ekstein, a Hasidic rabbi in Brooklyn, New York, devised a testing method that brought the Orthodox into the DNA age, although it was not the way Ostrer and his colleagues would have wished. Reclusive and snappish, Ekstein had firsthand knowledge of Tay-Sachs disease. He and his wife were carriers. Their first two children died of the disease before the age of four. Exceeding the odds, the couple produced a total of four sick children within a brood of ten. In shame, Ekstein had hidden his fourth doomed child from view; after the boy died, the rabbi was all the more ashamed for having done so.

An expert on the intricate dos and don'ts of *kashrut*, the Jewish dietary laws, Josef Ekstein took it upon himself to study Mendel's laws. Genetics, after all, was about applying the letter of the law of nature. The rabbi believed that God had created an order, a natural course of events, which scientists and doctors ought to investigate. Everything is managed from upstairs, Ekstein explained in his gravelly voice, but nature also has its rules. As before the great Flood, when the subdivisions of nature, clean and unclean, two by two, marched up the ramp beneath Noah's discriminating gaze, this one goes with that but never with this, the wonderful hairsplitting order of life, of which there was no more wonderful example than the biochemical rule pairing A with T but never with C or G, and the complementary rule matching G with C but never with T or A.

Ekstein's idea was to dissuade the Tay-Sachs carriers in his community from getting together in the first place. Under his program, high school boys and girls submitted their blood for testing, but the results were withheld from them. Later, when a pair started to date in earnest or their families were exploring a match, they contacted the database to learn whether they were genetically compatible. Most of the time they were. If incompatible, they were strongly urged to part and find someone else. Everything was handled confidentially—no names were recorded, only an identifying number and birth date for each youth. Ekstein called the program Dor Yeshorim, meaning Generation of the Righteous. By the late 1980s, the decline in Tay-Sachs births was more dramatic within Orthodox neighborhoods than outside of them.

Dor Yeshorim occupies a small, graffiti-splashed brownstone just south of the rumbling Williamsburg Bridge, near Brooklyn's East River waterfront. On the first floor of the office, half a dozen young women, all wearing head scarves, fielded phone calls at computer terminals. It looked like a telemarketing operation that had taken a deadly serious turn. We took the medical science and we applied it to the needs of the community, Ekstein said, gesturing proudly.

He believed he had been born a Tay-Sachs carrier for a reason. I have a mission to prevent misery, the rabbi explained. I had to find out the hard way, but we think everything happens for a divine purpose. We can override, with good deeds, the bad things in the universe. If you're a carrier of a disease, he said, it's predestined, but it's also for a purpose. You were also given the means to protect yourself. It is like clothing against the weather. You have to protect yourself with whatever tool you're given by the Almighty.

Ekstein went after the other recessive diseases affecting his community. Although it was not his intention, he was one of the instigators of the DNA era because he started looking for disease genes just as biotechnology became capable of discovering them. There are some forty Ashkenazi genetic disorders, each very rare, but up to half of Ashkenazim harbor at least one adverse mutation. The mutations act like the submerged mass of an iceberg. Absent prevention, about one in three hundred Ashkenazi matings will produce a child with a genetic defect. Ten recessive conditions, less rare and more serious than the others, deserve mention. Tay-Sachs, Niemann-Pick, Canavan, and mucolipidosis type IV top the list because they are essentially untreatable and usually lethal in early childhood. Bloom syndrome, familial dysautonomia, Fanconi anemia, glycogen storage disease, maple syrup urine disease, and cystic fibrosis leave the young patients gravely impaired. A battery of drugs can prolong their lives but do little to alleviate their suffering. Cystic fibrosis, to be sure, affects all Caucasians; it is the one disease in the group for which Ashkenazim aren't at higher risk. The core list of conditions rounds out with Gaucher disease, which has the most carriers but also is the easiest to treat. Again, it should be stressed that all these illnesses crop up in other populations, but at lower rates.

Focusing on the top ten, Josef Ekstein added carrier tests to his program as the underlying genes became accessible. In effect, he expanded the grounds for incompatibility among Orthodox young people. Still, ninety-nine of one hundred pairs received Dor Yeshorim's blessing to proceed with courtship or dating, their carrier profiles unrevealed to them. Meanwhile, the hundreds of thousands of blood samples that Dor Yeshorim gathered have been a great resource for DNA research. Dor Yeshorim contributed to the discovery or the confirmation of the genes for Gaucher disease, Canavan disease, and Fanconi anemia. In the late 1990s, during a race to discover the gene for familial dysautonomia, Ekstein and his scientific collaborators competed with a rival group that included Harry Ostrer. That didn't help the relationship between the two men, which was already strained by their opposing views about genetic information.

The rabbi would spare Jews the psychological burden of their DNA so that they would focus on making a good marriage and a healthy family.

They would be fruitful and multiply as the Almighty had commanded, while he, Ekstein, worked to rid the community of genetic disease. But for Harry, a Reform Jew, a scientist, a liberal, the primary commandment was informed consent. He wanted to eradicate the Ashkenazi disorders as much as the other man did, but not through an authoritarian and secretive program. Rabbi Ekstein wants to control who does and doesn't get information, Harry said. As you know, that is contrary to the egalitarian and participatory style of genetic counseling.

Genetic counseling of that sort, said Ekstein, is not there to take away the worry but to *increase* the worry. The rabbi nursed a broad critique of modern medicine. Having worked with a good many doctors and scientists over the years, he had decided that the professionals made it more difficult for a man and a woman having a baby, not easier. There was too much unnecessary testing, too much technical information being generated that was not supportive of the persons who had generated it. Through the tests and scans the doctor offered parents a sense of control over the birth and health of their child, but he caused tremendous anxiety in the interim and in the end it wasn't his problem. The doctor stood there with a printout from the lab. What did the couple want to do, considering the odds? The science of genetics is a tremendous tool, it can be very good, Ekstein stated, but most doctors are not using it the right way. They lack compassion for their patients.

This was a stiff charge if he had Harry Ostrer in mind. Harry hated to hear of congenitally damaged children. After a residency in pediatrics he had gravitated toward the academic aspects of medical genetics in part because the clinic was painful for him. In medical school I wanted to be a pediatric oncologist for a while, he said. Then I found out: They all die. (Prognoses are much better today.) Harry's global interest in the DNA of Jewish populations sprang from a sad desire never again to hold the hand of a distraught Jewish parent. At New York University Medical Center, a sprawling complex in Manhattan across the East River from Ekstein's redoubt, Harry offered genetic testing to Jews and non-Jews alike, about five thousand patients per year. Many of the patients were pregnant, educated white women who were concerned about something going wrong and who wanted answers fast. The lab's eight genetic counselors provided information and referrals before and after the testing.

The information of course could be upsetting. But Harry believed that if a person wanted a test, a glimpse of the future for herself or her fetus, it should be provided, just as long as the patient understood the implications. Consumers don't want to be told what to do, he said. Moreover, since other medical centers in New York promoted the same genetic services, Ostrer, to stay in the forefront, pushed to expand the number of tests that NYU offered. We are adding value to patient care—it is a growth business, he said. Let the marketplace decide.

But Ostrer wholeheartedly agreed with Ekstein that a couple shouldn't be exploring their genetics during pregnancy, which was already a stressful time. Carrier status should be factored in much sooner. So periodically Harry went to college campuses to screen Jewish students, intending to intervene before they married, as Ekstein did. On religious campuses such as New York's Yeshiva University, Ostrer and his helpers competed directly with Dor Yeshorim to raise student awareness of the recessive disorders. Personally drawing the boys' and girls' blood kept him connected to his original mission as a doctor. I am trying to prevent Tay-Sachs disease, Harry said stoutly. People are still falling through the cracks. The modern Orthodox kids, he added tartly, want to know their results. Like Dor Yeshorim, he saved the blood samples for his own research, although for that part of the work the subjects signed consent forms stipulating that their names would be stripped from the samples.

There was a political element to the information. Outside of their own forums, websites, and newspapers, Jewish health professionals have not courted publicity for genetic screening. On the one hand, they didn't want to reawaken early twentieth-century stereotypes about sickly Jews. On the other hand, they were sensitive to the charge that mass screening was akin to eugenics, a subject with Nazi overtones. Ekstein in particular was accused of practicing eugenics. Several years ago Ostrer and a colleague, Susan Gross, lobbied Jewish foundations and philanthropists to underwrite a national program of carrier testing and genetic education about the recessive conditions. Would organized Judaism, in

Harry's words, rise to the occasion? It didn't back the campaign, and Ekstein for one was just as glad. All this attention on Jews! he exclaimed. More recently, a group called the Jewish Genetic Disease Consortium—Harry was one of its medical advisers—has taken a higher profile. The organization has enlisted rabbis in its educational effort.

In the DNA age, no group of human beings has undergone a greater genetic scrutiny than Ashkenazi Jews. A 2003 study found that in relation to the size of their population, Jews were overrepresented in the human genetic literature. A preponderance of Jewish geneticists is one reason—the positive-feedback loop. Another reason is that Ashkenazi DNA is relatively easy to study, because there's less clutter in the genetic structure, owing to the group's history of inwardness. A third reason is that Jews readily volunteer for research; many regard it as a divine command, a *mitzvah*, to participate. While other minorities, such as blacks and Native Americans, backed away from genetic probes, fearing that the results could be used to discriminate against them, Jews took custody of their DNA, both the raw material and the interpretations.

In short, the community was well prepared—in its fractious fashion, ranging from Ostrer's camp to Ekstein's—for the revelation of BRCA1.185delAG in the mid-1990s.

Whenever a field of natural science produces a swarm of noteworthy results, as genetics did during that period, you can expect to find great gains in the technical apparatus running ahead of the field and laying the groundwork. Because the text of the human genome was both very long and very small, a complex set of editing and magnifying tools was needed to make sense of it. Hence the development of restriction enzymes, which cut DNA molecules into fragments; ligases, which glued DNA back together; cloning, which inserted pieces of DNA into colonies of bacteria or yeast and vastly ramped up the volumes; PCR, an ingenious system for fishing out and copying pieces of DNA; then the sequencers and microarrays to read the letters of the DNA fragments; and finally, down the hall from the wet lab, bioinformatics, the software to wrangle the herds of data through computers. Its four-letter alphabet easily digitized, DNA was a natural for computer programming. When computer science was applied to molecular biology, a new specialty was created, called genomics, standing for the lightning-fast appraisal of multiple genes. Gene, genome, genomics: that was the cutting-edge progression.

If you ask people to name the greatest triumph of the DNA age, they probably will say the high-tech human genome project, not the discovery of BRCA1 and 2, the last great genes from the simpler world of Mendel. With each year that passes, the argument looks better for reversing that judgment, because BRCA has had a large impact on medical decision-making in the United States, and the genome project feels to have been hyped. The project, which began in 1990, took the form of a competition between government and private research teams to be first to inventory all three billion letters of human DNA. Construed as both a catalog and a map, the project aimed to guide scientists through the four-letter wilderness of the genome to places of medical value. (The genes themselves occupy only 2 percent of the genome, with the function of the rest unclear. Most DNA is extraneous, if not useless.) When the two research teams declared a tie and published the human DNA sequence in 2001, there was much cap-flinging and commentary about the impending transformation of medicine, along with some nervousness about what the knowledge would mean for individual privacy and for racial and ethnic identities.

To talk about the human genome, singular, is somewhat misleading. There is not one human sequence but billions, each highly similar, each slightly different. The variation in human genomes underlies not only people's physical diversity but also their different vulnerabilities to disease. DNA variation explains why, under equal conditions, some people will be disposed to high blood pressure, diabetes, asthma, or heart disease, and some people won't. Diet, lifestyle, and landscape—the environmental variants of health—have critical parts to play in the disease process, but these familiar factors were outside the bailiwick of genomic technology and to science journalists weren't nearly as exciting as the new information spewing forth from DNA.

Understanding the so-called complex disorders of medicine was the ultimate purpose of the human genome project. Previously, researchers had been nickel-and-diming their way through the DNA. They had netted genes for the rare mendelian conditions, such as Huntington's disease, cystic fibrosis, and the Jewish genetic disorders. Collectively these were orphan diseases because there weren't enough patients to warrant big investments by pharmaceutical companies. The profits lay in the complex disorders and their large cohorts of patients. With the advent of genomics, the search for single genes having a big impact on a relatively few people shifted to multiple genes having a modest impact on a lot of people.

Where did breast cancer fit in? Breast cancer straddled the old and new paradigms in genetics, the mendelian or single-gene model versus that of genomics. With two hundred thousand new cases in the United States each year and forty thousand deaths, breast cancer was far from an orphan disease. Breast cancer's incidence was high, yet its pattern was scattershot and its causes were murky. Attention fell upon the roughly 10 percent of cases that clustered in families. The inheritance pointed to a mutation operating in a dominant, not recessive, mode. Here, the silent possibility of healthy carriers didn't seem to apply. Although the unknown gene could not take in all of breast cancer, the pathway it followed in the body might provide a handle on the sporadic disease. Identify the gene, already dubbed BRCA, and you might hook into the cancer pathway and learn how to interrupt the broader condition. If the transition between the eras of Mendel and genomics was going to succeed, breast cancer was the bridge, or so it was hoped. The drug companies were keen for it.

There turned out to be two BRCA genes. The first was identified in 1994 and the second the next year. The subtext of the hunt for the genes was the snippy rivalry between Mary-Claire King, then at Berkeley and now at the University of Washington, and Mark Skolnick of Myriad Genetics in Salt Lake City. Skolnick was a breezy scientist. In an interview he referred to himself as a Jewish kid from New York joking around with Mormons. He had found that Mormons made good subjects for DNA studies, thanks to their large families and their thorough genealogical records. As for Mary-Claire King, well, she was not breezy and not Jewish. A decade later, King revealed that her grandfather was a Jew and that her mother had concealed the fact because of antisemitism when she was growing up.

Both researchers had collected pedigrees of heritable breast and ovarian cancer since the 1970s. King, taking the lead, showed that chromosome 17 contained what would be the BRCA1 gene. Knowing the chromosome was akin to knowing the street where a suspect lived. For months the two research teams worked the street up and down, looking for BRCA's molecular address. Most of the technology they used was pre-genomics and thus rather slow. But it was Skolnick's lab that won the race to publish the sequences of BRCA1 and BRCA2, and Skolnick's company that became rich from patenting the BRCA genes and marketing tests for them. The assiduous King, with a string of accomplishments before and after, earned the greater reputation in genetics.

Since BRCA1, as has been noted, is a very long gene, thousands of possible misspellings can occur along its sequence, and in fact more than two thousand mutations have been recorded to date, affecting all ethnic groups. One mutation quickly emerged ahead of the others. Although Skolnick and King hadn't keyed on Ashkenazi Jews in particular, other researchers, now that they knew where to look, noticed the prominence of the 185delAG mutation in their study samples. They immediately established its Jewish tie. Those samples had come from Ashkenazi families with a lot of cancer.

The next question was, how prevalent was the mutation in the general population of Jews? Researchers went back and looked for the mutation in DNA that had been stored from the Tay-Sachs and cystic fibrosis screening programs a generation earlier. The prevalence of 185delAG was 1 percent—meaning that one of a hundred Ashkenazim was walking around with a gene that could kill by itself, unlike the recessive mutations. Commenting on the surprisingly high frequency of the mutation, Francis Collins, the government's top medical geneticist, said that the breast and ovarian cancers attributable to 185delAG might be the most common, serious, single-gene disease yet identified in any population group.

A less-frequent BRCA1 mutation, called 5382insC, was identified next, followed by 6174delT on the other gene, BRCA2, making
for a trio of characteristically Jewish breast-cancer mutations. They had
stemmed from three different founders in Jewish history. There were
hardly any other BRCA mutations in the group to speak of. In most
other ethnic groups, the prevalence of BRCA mutations is low, about
one in four hundred. On the basis of their small set of three, the BRCA
carrier rate among Ashkenazi Jewish men and women is one in forty, ten
times higher than the background rate. Comprising only 2 percent of
the American population, Ashkenazim learned they represented up to a
third of the country's BRCA carriers.

Being the oldest of the three mutations and the first to enter Jewish
consciousness, 185delAG became the most studied, most worried-over
piece of DNA in the world. Largely because of it, more people have been
tested for BRCA1 than for any other human gene. The initial screening of the Jewish population, conducted by the National Institutes of
Health (NIH) in 1996, followed the procedure that had succeeded with
Tay-Sachs testing. Some five thousand volunteers in the Washington,
DC, area were recruited with the aid of rabbis and community health
officials. They provided researchers with DNA samples and histories of
any cancers in their families. As before, the people who stepped forward
to have their blood taken were relatively liberal and well educated, and
they offered to help even though, in this first snapshot of the problem,
they were not given their individual test results.

From the NIH study and others, scientists determined that not every
BRCA carrier was going to be affected, for the mutations were not fully
penetrant. Penetrance is a technical term signifying that a powerful gene
may pull its punches. Because of incomplete penetrance—because of
healthy, elderly carriers like Shonnie's grandmother, Dorothy—the risk
of cancer from 185delAG and other BRCA mutations must be expressed
as a probability. As genetic counselor Jeffrey Shaw had explained to the
Medina-Martinez family, the estimates are wide—the most hopeful
studies suggesting that just half the carriers will get breast cancer during
their lifetimes, the least hopeful that 90 percent will. The penetrance of

ovarian cancer is lower, in the neighborhood of 40 percent. Something blunts the thrust of the gene, but no female who is a carrier can count on that happening.

Since Ashkenazi women have more BRCA than other ethnic groups, while being just as susceptible to sporadic or "regular" cancers, they might reasonably be expected to have more breast and ovarian disease overall. The question could be resolved if scientists were to make the effort, but they haven't chosen to study it comprehensively. If there *is* an incremental impact on the group from BRCA, it is relatively small.

The impact of the mutations has been easier to measure in Jewish families where breast or ovarian cancer already was an issue. If a woman was ill, what was the likelihood that a mutation was the reason? The odds of a Jewish woman's being a BRCA carrier were especially strong if she had ovarian cancer or if other members of her family had cancer too. Another tipoff was the early onset of a breast tumor—say, in a woman in her thirties. In the late 1990s, a rule of thumb was put forward in the medical literature: the younger the Jewish patient with breast cancer, the more likely she would test positive for 185delAG. This finding in Jews influenced the official name of the newfound gene: breast cancer #1, early onset.

In 1998, hospitalized in San Diego and close to the end of her life, Shonnie was advised to have a BRCA test. Her age, just twenty-six when the tumor appeared, must have rung a bell with someone at last. Joseph and Marianne went into their depleted savings and paid $2,800 for the analysis, which was not covered by her insurance.

Hmm, said the genetic counselor, your daughter is a carrier of 185delAG . . . but that's the Ashkenazi mutation. The counselor must have been puzzled indeed. Given all that was going on with Shonnie, they put it out of their minds. The test was of no use to the patient—it was a warning to the rest of the family, but either the information wasn't delivered properly or didn't sink in.

When the news of the Ashkenazi BRCA was conveyed to the Jewish community, it was oversimplified in the direction of alarm. Women tended to overestimate their risks of carrying a mutation if there were any reports or hints of breast cancer among their relatives. Newspapers and magazines published articles by or about Ashkenazi women agonizing over BRCA. According to surveys, Jewish women were much more likely to visit a genetic counselor than other minorities were. The women knew that if they took the test and it was positive, the next dilemma would be whether to have a preventive mastectomy or to try to circumvent the risk with extra mammograms and closer surveillance. The knowledge was like the apple that Eve offered Adam. A single bite of the apple tempted another, as sisters, aunts, fathers, and daughters submitted their DNA for analysis. Positive results brought fear and shame, and negative results relief and guilt, the riptides of emotion dividing families.

Pressing to save lives, some scientists recommended large-scale screening of the Ashkenazi population, while others said no, it's too soon, better wait for additional studies to be completed. But how would such studies be used? Would health-insurance companies call BRCA a preexisting condition and deny coverage? There were laws against genetic discrimination in health policies, but they didn't apply to life-insurance policies. Would Jews become an uninsurable class of people?

Orthodox Jews wrestled with all of these anxieties and more. Within their circle, a diagnosis of breast cancer could prompt a woman to seek treatment in another city, lest the people at home find out. High in the patient's mind was her children's marital prospects. If a young person learned that a BRCA mutation had been detected in her family or, G_d forbid, that she carried a mutation herself, at what point in her courtship was she obliged to disclose the information to her possible spouse? Should she then be rejected? Rabbi Ekstein, unable to devise a workaround for couples in his care, found the situation deplorable. What is the advantage of screening for a condition, a mutation, that has no treatment? he demanded to know. It is just making more tragedy. For the rest of her life and her future generations, she will be burdened with worry. Ekstein came out strongly against the BRCA test.

DNA crossed the millennium under something of a cloud. From the beginning of the human genome project, a contentious body of literature and commentary had gathered around the ethical, legal, and social implications of genetics (ELSI for short). As with no other national research endeavor, the government funded sociologists, psychologists, and ethicists to look over the shoulders of scientists. The vetting was problematic, because the ELSI scholars fanned the very fears they were asked to assess, publishing warnings about the commercialization of genes, about discrimination and ethnic stereotyping, and about the manipulation of human traits under the guise of DNA repair.

The stalking horse for these dire scenarios was BRCA and its notorious mutation, 185delAG. In reaction, some Jewish leaders advised people not to participate in genetic studies any longer. This is what we get for helping? they asked, linking concerns about Jewish genetic defects to Nazi racial classifications and the Holocaust. To be fair, the ELSI scholars did not play upon the Holocaust. Still, with very few exceptions, the genetic abuses that critics warned about haven't materialized, and the dehumanizing scenarios remain very far in the future. The courts are the place to look for evidence about discrimination. Since BRCA was discovered, there have been no court cases documenting discrimination by health insurers or employers over a person's BRCA result.

One of the better works of ELSI scholarship was "Ashkenazi Jews and Breast Cancer: The Consequences of Linking Ethnic Identity to Genetic Disease," published in 2006. The authors, social scientists at Columbia University, complained that to target Jews as a group for the study of breast-cancer mutations was undesirable and perhaps unwarranted. Other groups had BRCA mutations that weren't being examined to the same degree, they wrote; hence their medical needs might be neglected. The authors pointed out that two of the three so-called Ashkenazi mutations had turned up in populations that weren't Jewish, 185delAG especially. It was careless reasoning, the paper reiterated, to brand the mutations Jewish and wrong to saddle Jews with BRCA while underserving others in the same predicament.

BRCA researchers did not dispute that they sampled Jewish women disproportionately, but as Harry Ostrer and several of his colleagues observed, it made good scientific sense to explore the nature of BRCA through Ashkenazi DNA. Quite apart from their willingness to cooperate, Jewish subjects were efficient to study. It made sense to solicit DNA from a group that you knew, going in, contained five or ten times as many BRCA carriers as any other you might choose. An enriched population greatly reduces the number of DNA tests that have to be performed on a hard-won collection of patients, not to mention the savings in hours of genetic counseling.

An example of the approach was the New York Breast Cancer Study, directed by Mary-Claire King. With assistance from Ostrer and a dozen New York area investigators, King pulled together a group of one thousand Ashkenazi women with breast cancer. Ten percent duly tested positive for BRCA. From the one hundred carriers, King scaled up the sample to encompass seven hundred of the carriers' relatives. She tested their genetic status and recorded their health histories, whether they had cancer or not. From the enlarged pool of families, she calculated that the penetrance (the proportion of carriers who got the disease) of BRCA was 82 percent, a high-end projection for breast-cancer risk. There was somewhat happier news in King's finding that physical exercise and staying thin could help a younger carrier forestall the disease. Because the statistical analysis was robust—biomedical science really appreciates large sample sizes—the results of the New York Breast Cancer Study, published in *Science* in 2003, have been cited many times in the literature and the mainstream media.

Not to be overlooked was the cost benefit of working with Jewish subjects. Protected by its patents, Myriad Genetics sets high fees for sequencing the BRCA genes, charging around $3,000 per test for a consumer and half that for researchers. They have a stranglehold on the test, Harry groused. He had to ship his samples to Myriad in Utah when his own lab was perfectly capable of running the analyses. But the company charged only $400 to read a Jewish subject's sample, since it looked for only three mutations. The sequencing zeroes in on just those two sites on the BRCA1 gene and the one site on BRCA2 where a misspelling might occur, rather than scanning the full length of the genes and searching for any number of mutations, a procedure that obviously is more expensive. The discrepancy between what a Jewish woman (or her insurance company) had to pay versus a non-Jewish woman's cost was part of the reason why screening lagged in other communities. Access to care—the rich and well-informed able to buy genetic services, the poor not—was a bona fide issue for ELSI.

Like it or not, the Jewish BRCA carriers were the vanguard. Thanks to the 185delAG mutation and its two companions, BRCA has become a metonym of cancer risk for every woman. A genetic aberration conceived in isolation from the rest of the world was donated to the world through science, adding to knowledge of BRCA's universal behavior. If there was one disadvantage to using Ashkenazim in genetic studies, it was that their blood relatives were harder to come by than in other groups. Their extended families tended to be smaller, their pedigrees pinched off, because of the Jewish losses in pogroms and the Holocaust—a deficit that Rabbi Ekstein was trying to correct by encouraging healthy breeding, even as he complained about the harmfulness of BRCA testing to families.

In New York, Chicago, Los Angeles, and other cities, demand for the tests surged, in part because Myriad Genetics promoted the service directly to doctors and consumers. The company reported in 2008 that the number of people taking the test was growing by 50 percent every year. The negative findings well outstripped the positives, but that didn't dent the demand to know. Myriad estimated that half a million BRCA carriers in the United States remained to be detected, including 150,000 who were Jewish. Among five million American Ashkenazim you could figure on fifty thousand carriers of 185delAG. To be sure, nobody in the genetics profession was calling 185delAG an Ashkenazi mutation any longer, that is, specific to that group of Jews, because by this time it had turned up in a dozen distant places along the spidery trail of Jewish migration—as in the ex-Kingdom of New Mexico.

After Jeffrey Shaw's counseling session for the Medina-Martinez clan in the summer of 2007, three of the Medina sisters, Wanda, Lupita, and Louisa, along with their mother, Dorothy, and Wanda's husband, Bill, went to Colorado Springs to sign up for a 185delAG test. Thanks to a state grant, the cost to them would be minimal. Joseph Medina and Iona meant to go too, but they didn't make it. A mix-up in communication or something.

Shaw remembers that everyone was in good spirits. They weren't worried about the gene, he said. They weren't worried about the, quote, little bit of cancer, unquote, they might have in them already. No gloom and doom. They talked about cleansing their livers with diet and resetting their bodies. When the DNA results came back from Myriad, Dorothy (the obligate carrier) and Wanda and Lupita were positive for the mutation. Louisa was negative. Three more positives atop Shonnie's positive and Chavela's positive.

Bill and Wanda had never doubted that Wanda already was battling breast cancer, since a lump had appeared on her ultrasound scan earlier that year. They did not want a biopsy. Bill and a local nutritionist, who was, like the others, a Jehovah's Witness, took over Wanda's care. For many months everyone believed that the special diet was working.

In March 2009, Wanda was hospitalized with end-stage metastatic breast cancer. A hurried mastectomy came too late, and she died in June, at the age of forty-seven. In the weeks before the end, she tried frantically to rearrange her life.

The educated medical consumer was used to the notion of risk factors for disease. He or she tended to regard them as painless early warning signs. To live in a state of risk wasn't alarming when risk factors like eating and exercise habits could be met head-on. In principle, you could control your health environment, although the informed medical consumer also knew that diet and exercise went only so far.

When your doctor spoke of your probability of illness, the risk was said to be high or low, a relative risk, that is, compared with a murky average, not your personal numerical odds. Supporting your risk profile was an array of screening tools, such as cholesterol measurements, blood-pressure tracking, and various scans, whose findings might elevate your risk, but usually drugs could be prescribed to drive it back down. The situation was manageable. If risk was like a clock, the patient figured she still had plenty of time.

DNA analysis pushed back the start of the risk–to–disease continuum. When DNA was added to the risk calculation, through a session with a genetic counselor at a medical center or through a commercial testing service, the patient might discover that her clock for one or more conditions had been ticking for much longer than she knew, and that the vague risk factor known as family history had converted to a hard number, a probability with her name on it. At that point, a lump rising in her throat, she was likely to go for additional testing and to consult one or more specialists. The collusion between American medical technology and the patient's empowerment—the self-affirming need to gather as much information as possible in order to have a strong voice in her care—might shift against her. Now when she stepped onto the conveyor belt of monitoring, instead of reassurance she got glitches, benign spots and masses, equivocal readings, the nerve-racking false positives, which only delayed the day of diagnostic reckoning. The 185delAG carrier has pioneered the way into the risk–disease limbo, and many more people, the potentially sick and the worried well, will follow for other conditions.

Doctors are unhappy when they can warn but not treat. The prospect of a cancer coming sooner or later was extremely stressful to their BRCA patients, as doctors appreciated. To spare women both the anxiety and the jeopardy, prophylactic surgery is recommended: bilateral mastectomy to prevent breast cancer, and for ovarian cancer, bilateral salpingo-oophorectomy, which removes the two fallopian tubes as well as the two ovaries. In the mid-1990s, when BRCA was new, cutting out the problem was the nuclear option, and patients shrank from it, but today doctors can show statistical proof that the operations sharply reduce the occurrence of tumors. Although breast cancer can be treated success-fully, ovarian cancer is a bad cancer to have, since 70 percent of the cases are advanced by the time the doctor sees them. If a carrier insists on preserving her breasts, ovarian removal alone should lower her chances of breast cancer, because of the elimination of estrogen from her system. If she wants children, the ovarian surgery can wait until she is through childbearing. Encouraged to look past the trauma, the patient could get back to normal with surgically reconstructed breasts and without the reproductive organs she no longer needed.

The advice, so very rational, took hold. According to an international survey published in 2008, American BRCA carriers had the highest rate of bilateral mastectomy in the world, about a third of U.S. women choosing to have preventive surgery and about two-thirds of the carriers also having the ovarian procedure. BRCA helped to fuel a trend, as even noncarriers with breast tumors were choosing to have the other breast removed.

Yes, knowledge was power. Women did their research and shared notes about their surgeries. There was a righteous solidarity in the name that mutation carriers picked for one another: previvors, meaning the survivors of a predisposition to cancer. The name came out of an online organization called FORCE (Facing Our Risk of Cancer Empowered), which was founded in 2000 by an Ashkenazi carrier. More recently, when the blogosphere took up the issue, you could follow The Secret Life of a BRCA1 Mutant, in which a 185delAG carrier told of the events surrounding her pbm (prophylactic bilateral mastectomy). Her blog was not to be confused with I'm a Mutant: The Life and Times of a BRCA1 Previvor, by another 185delAG carrier who was looking forward (not!) to her surgically induced menopause. The bloggers at Cut the B*tches Off and Goodbye to Boobs, respectively, worked themselves into a cold fury over the logic of slice-and-gut (slice the breasts, gut the ovaries).

Collectively, as their blogroll put it, they were the Sisterhood of the Mutant Bloggers, defiantly complying with their medical advisers.

The previvors included more than Jewish women, but all were regarded with pity by Rabbi Josef Ekstein. Remove the breasts, remove the ovaries? he grumbled. Next they will say to remove the brain. . . . I'm against BRCA testing, Ekstein emphasized. Why test? To prevent a disease? To heal a disease? Or to put a stamp on a person that she has the gene? It does more damage than good—you label the family, you *kill* the family, he said, for a moment sounding overwrought. He liked to quote King Solomon's declaration in Ecclesiastes: More knowledge, more pain. This is not to say, Ekstein elaborated in a written commentary, that one should be a fool. However, although Solomon was the wisest of all men, he recognized that in certain instances it is more helpful not to be informed.

Occasionally when the rabbi ventured out of Brooklyn to lecture, genetic counselors would press him to justify his position. His critics stood strongly for science, which shaved away the uncertainty of nature in slices, as a CT scanner does, and advanced patiently toward the day when all things would be known. In the meantime medical information should be transparent and every patient was autonomous. Ekstein faced the questioning politely—a roly-poly, yarmulked figure with a shaggy white beard and brows as dark as thunderclouds—and he gave not an inch. He had nothing against science, he said, only its hurtful applications to those who were psychologically vulnerable. He flashed a slide on the screen: EXPERIENCE SHOWS THAT EVEN THE MOST HIGHLY EDUCATED OF INDIVIDUALS HAVE DIFFICULTY USING INTELLECT TO DOMINATE EMOTIONS.

When pressed further, Ekstein put forward a series of arguments, the effect of which was to compound the uncertainties of BRCA testing and undermine its medical utility. First, even if she weren't a carrier, a woman had a one-in-eight chance of getting breast cancer during her lifetime, owing to the sporadic or background risk. Second, genes other than BRCA, including genes still unknown to doctors, were responsible for some of the heritable cases. Third, because of BRCA's incomplete penetrance, its faulty guarantee, as it were, a significant portion of the women who had mutations would not contract the disease. Fourth, turning now to the surgery, even that harsh measure could not completely preserve their health, since malignant cells were sometimes missed.

Ekstein advised that women who were worried about BRCA should step up their vigilance and be screened more often, by MRI as well as by ultrasound and X-rays. If her family history indicated she was at risk, she must take all the precautions of breast health—things she should be doing anyway, Ekstein said, without the need of a test, without putting a blot on her relatives and living her life in terror.

All in all, the rabbi made a forceful presentation. Except that he ignored the poison pill of ovarian cancer, a fifty-fifty probability, roughly speaking, for the mutation carriers who chose to preserve their ovaries. The risk of ovarian cancer was what frightened the doctors and counselors, and any potential BRCA carrier, Jewish or not, ought to be aware of the danger, since neither the screening methods nor the treatments for ovarian cancer were reliable, while preventive surgery has been shown to cut mortality dramatically. Perhaps that is why, when he finished defending his position, Ekstein made a slight bow and said, We cannot set a general rule for the whole world.

Coming to grips with their BRCA risk, some Ashkenazi professionals have asked again if they should have a national screening program, patterned on the old Tay-Sachs testing model. While Canada, Britain, and Israel started programs offering BRCA testing to Ashkenazi citizens regardless of family history, the U.S. Jewish community cautiously aired the pros and cons. On the one hand, according to a study by Wendy Rubinstein, a BRCA expert in Chicago, a moderately successful screening program would save about five thousand lives on the ovarian-cancer front, not to mention the lives threatened by breast cancer. The most ambitious scenario was that the three deadly mutations would be expunged from the Ashkenazi population via adult screening, prenatal genetic diagnosis, and termination of fetal carriers.

On the other hand, Harry Ostrer asked, why should Jews be the forerunners, the elites, in this effort? Qualms from his quarter were surprising, given that Ostrer worked to stamp out the recessive Jewish disorders through communal screening. I'm for equal-opportunity testing, Harry explained, not just for 2 percent of the population. I mean for everybody else. Gentiles should be offered BRCA screening. I could develop good tests for the WASPs of Connecticut [i.e., tests for their mutations], but I'm not empowered to do that.

Ah, the Myriad patents again, crimping independent research on the two genes. In one of his NYU research projects, Harry was looking at the question of penetrance—specifically, why some Ashkenazi women with BRCA mutations didn't get cancer—and the research would be cheaper if he didn't have to pay Myriad for sequencing the two genes. In fact, Harry had joined a class action against Myriad over the patent issue. Though a district court ruled favorably, the lawsuit, he admitted, would be a long shot in the higher courts because of the precedents supporting the company. It did seem wrong that a piece of nature could be patented commercially, but for thirty years courts have upheld the principle, not so much on the ownership of genes as on their transformation into biomedical tools.

Ostrer also voiced a concern about widespread BRCA screening that made him sound a bit like his rival, Rabbi Ekstein. Some have called this no-brainer testing, Harry said at a symposium. But is this true? Genetic testing without adequate genetic counseling can lead to fear and anxiety. Indeed, there were even instances, as Ostrer knew well and Ekstein knew better, in which a woman with a positive result had dropped out of the medical system altogether, refusing to return calls from her genetic counselor or doctor, so frightened was she of finding a tumor. Those people took no comfort in being previvors. They were better off not knowing, yet they knew.

Another Medina sister, Lupita, having tested positive for the mutation in 2007, was diagnosed with breast cancer in 2009. She opted to have a bilateral mastectomy.

Shonnie's, Wanda's, and now Lupita's cancers made it impossible for

Iona to put off her BRCA test any longer. She was having trouble sleep-ing. I realized I was being selfish, Iona said. It would be selfish of me to put people through what Wanda and Shonnie did, if I could do some-thing about it.

Leaving a blood sample at the lab, Iona went home to await the result. Her heart started beating rapidly. To work off steam, she did kickboxing exercises against a heavy punching bag, and one day she felt a stabbing pain in her chest and was taken to the emergency room, believing she'd had a heart attack. Instead, she had pulled a muscle.

Then the genetic counselor phoned: Iona, like her late sister and at least ten of her close relatives, was positive for 185delAG.

In hindsight BRCA1 and 2 were the last truly important human genes to have been discovered. Although many subtle findings have been extracted from the human genome since the mid-1990s, they are a large cast of bit players making small impacts on health and illness. The new-found DNA markers and genetic variants contribute tiny, incremen-tal pluses and minuses to the risk of disease, of negligible importance to the individual patient. Almost certainly there are no undiscovered stars left in the genome, no genes that might be given makeovers by the drug companies and brought forward to cure the common illnesses of mankind.

Nor did BRCA itself provide insights into the general pathway of breast cancer, because it has become clear that breast cancer comprises several overlapping conditions. The kinds of tumors that BRCA carriers usually developed weren't like the tumors of common, sporadic cancers. The cell pathology was different, and the outcome for BRCA patients tended to be worse. However, targeted therapies were arriving. Much could be written about the new class of anticancer agents, which attack individual features of cancer cells instead of bludgeoning all cells indis-criminately, as conventional chemotherapies do. Genomic technology was helping to distinguish among the breast-cancer types; the typing of a tumor was the first step to tailoring a treatment. Genomic laboratories spewed As, Ts, Cs, and Gs into the cancer data banks as if from a fire-hose, for it was easier to gather the data than interpret it.

Ten years into the new century, the DNA age was moving sideways. A paradoxical sign that genetic science was adrift was that consumers were having fun with it. Recreational genomics—the specialists' dismissive phrase—was where the action was. Scientists watched uneasily as people explored their genealogies and their potential health risks through direct-to-consumer testing companies like 23andMe, Knome, and Navigenics. These services were accessible on-line. For several hundred dollars and a mailed-in swab of your cheek cells, you could see narrowly into your biological past or future.

Harry Ostrer championed the old-fashioned approach: You made an appointment with a medical geneticist or genetic counselor and took a DNA test when one was warranted, not just because you were curious. The consumer tests being marketed might be valid for aspects of ancestry, Harry believed, but not for a medical profile. I fear that the purveyors of direct-to-consumer genetic testing, with their faulty claims, will scare away our patients and betray the public's trust in medical genetics, he warned. Compared with the deliberative process that had introduced BRCA screening to the American public, the current gene scene was the Wild West. Harry and others called for federal regulation, and the Food and Drug Administration began to respond.

Lacking fresh insights, scientists brought the genome back into the shop and popped the hood. For all its power, the technology of genomics was still running too slowly. The mechanics worked to make the technology of DNA run faster while pushing down the cost of sequencing. Their hope was that one day everybody might be able to afford to have their genomes transcribed, all three billion letters' worth. Faddishness still seemed to drive the field. It was more about buzz than science when actress Glenn Close, aging rocker Ozzy Osbourne, and Archbishop Desmond Tutu had their genomes sequenced. A dozen or so individuals got together on a stage in Cambridge, Massachusetts, and described how they felt to be the first human beings to have their DNA read in full. It felt good.

———

When Iona found out that she carried the mutation, she promised to increase the frequency of her mammograms. Uninsured, she couldn't afford the cost. But in 2010 she enrolled in a new program at Penrose Cancer Center in Colorado Springs. It was called the High Risk Breast Clinic. The plan was to offer BRCA carriers a free consultation with a breast surgeon, along with mammography or an MRI scan, thanks to a grant from the local chapter of Susan G. Komen for the Cure. As it was, Iona hadn't had a mammogram in more than a year, a dangerous gap.

In a sense the medical circle had closed, for Susan Goodman Komen, in whose name the foundation was created, was a Jewish woman who died of breast cancer when she was thirty-six. She was Iona's age and a mutation carrier. Shonnie also had received financial aid from the Komen foundation.

Signing up for the new program made Iona feel better. Her first MRI was negative. In her mind, she had taken her breast- and ovarian-cancer risk under control through the strategy of watchful waiting, the strategy Rabbi Ekstein advised. Preventive mastectomy was off the table, and after learning more about the ovarian surgery, Iona decided she didn't want that operation either. The fact that she had taken birth-control pills for years was somewhat protective against ovarian cancer. Maybe when I'm fifty, she said about the oophorectomy.

The anxiety attacks, the sense of being overwhelmed, were past, but in relating this, Iona spoke promptly, as if she were answering a study question at the Kingdom Hall. More by rote than reflection, perhaps, and giving wide berth to her emotions. Experience shows that even the most highly educated of individuals have difficulty using intellect to dominate emotions. Iona knew the answer deep in her DNA, yes she knew. She knew what her headstrong sister hadn't, though Shonnie would not have cared.

Chapter 8

LAST DAYS OF THE INDIAN PRINCESS

On February 18, 1997, Shonnie Medina mailed a card to her sister in Texas. Iona and her new husband had been living in Odessa, his hometown, but they had recently decided to return to the San Luis Valley. Shonnie picked a Garfield the Cat greeting card. The exuberant cat flings his arms into the air, asking, Guess Who's Thinking of You? And inside, leaving plenty of room for her note: Me!

Iona: Hi there! So how are things going with you? Me, I'm excited for you guys to come. Hope you guys are too. I've been positive lately. I guess that's the way to be so we don't get discouraged huh? I'm sorry I never write to you. I always say that I'll get to it and I never do. Mom came over last week and spent the night, do you believe it! Just to come? (Of course, I had to ask her to.) ☺ *When you move here I know they will visit us more. I had to go to the clinic the other day because Michael got scared for me. I found a lump in my breast. I have to go to the Doctor on Monday. Man, if they have to take off one of my boobs, everybody will notice—Ha Ha!*

Anyways . . . everybody says hello and are excited for you guys to come. Hope you like it here. Tell everybody up there "hello" from Mike and I hope to hear from you soon. Call me to let me

know for sure when you guys are coming down. Thanks for being
such a great sis!

> *Love you lots,*
> *"Shaunee"*
> *Write Back Soon!*

The lump Shonnie detected in her right breast was about the size of a pea. Mom, she told her mother, I know, it's a gut feeling. It's cancer.

It's natural to fear cancer, even if you're only twenty-six, and the cancers on her father's side of the family were no secret to her. Still, Shonnie often had dark presentiments. People in Culebra remember her vivacity and poise, but her mother and sister knew she experienced bouts of dejection and jangled feelings, which she tried to keep private. One of Iona's small duties in life was to cajole her sister when she was down. That's what Shonnie had been referring to in the card when she said she'd been trying to stay positive lately. She was more emotional and moody than I am, Iona recalled. I'd balance her out. But if I'd go overboard on being positive in my own life, or being too timid, looking the other way and letting things go, I mean, Shonnie would say, Don't let people take advantage of you.

Dr. Gladys R——, a primary-care physician at Alamosa Family Medical Center, saw Shonnie on the same day that Shonnie sent the card to Iona. Since Shonnie was a new patient to R——, the doctor probably spent a minute reviewing her medical records. They are for the most part unremarkable, showing a dozen visits to the clinic since 1992 and covering the normal complaints and routine checkups of a healthy young married woman. Shonnie needed a birth-control prescription, a gynecological workup, and a couple of modifications to her birth control. She also requested treatment for allergies, colds, and acne. After February 1997, however, the case notes, test results, and medical opinion shift radically, revealing a breast-cancer patient at odds with her medical providers—so those early, unremarkable visits by Shonnie may merit a closer look. When a life is carved at the joints and cast onto a page, don't clues lie everywhere, pointing back to her DNA and rushing ahead to her end?

The very first record that the medical center filed for her, in March 1992, probably had to do with her being molested as a child. Engaged to be married in August of that year, a virgin who may have felt less than virginal, Shonnie reported being bothered by mood swings. She attributed them to a childhood incident that she didn't specify. She was referred to a counselor in Alamosa, and apparently she went to a session or two. The trouble came to a head a year or two later, in a confrontation with the offending uncle, and then, according to her family, Shonnie put the incident (or incidents) behind her. How much it colored her outlook is impossible to know. Her mother thinks this was why as a little girl she felt she would die young. That, plus the gene whispering in her blood, made her the way she was, Marianne maintained.

Complaining of red and itchy eyes, Shonnie appeared at the clinic in early 1996. Normally the nurse prepped her for the doctor by taking her temperature, blood pressure, and weight. But Shonnie refused to be weighed today. Since she'd been married, her weight had risen from 120 to 143 pounds, according to the records, and 143 was also her weight when her cancer odyssey started, a year later. At five feet seven, she'd gone from slim to statuesque. Marianne said that Shonnie obsessed over her weight. Marianne also would say, with nothing but love, that her daughter was vain. Shonnie was bothered by *any* imperfection in her body, however minor. Today, her eyes oozing from an allergic reaction of some sort, the young woman was in no mood to get on the scale, and she didn't.

A third entry is worth mentioning. Although the medical system did not handle her very well after February 1997, five years previously a staffer did teach Shonnie how to do a breast self-exam, without which she might have missed the lump until it became larger.

Shonnie's first move after detecting the lump had been to visit the walk-in clinic run by Planned Parenthood in Alamosa. The staff there thought it was a benign cyst, and Dr. R———, examining Shonnie the next day, reassured Shonnie that it was a cyst. The doctor described the cyst as single, mobile, and tender; a half-centimeter in diameter; and located in the lower, outer quadrant of the right breast. The doctor asked Shonnie to let her know if the cyst got any larger or did not shrink in a few weeks.

When Shonnie returned five weeks later, the cyst had doubled in diameter. R—— tried to draw some fluid out of it with a syringe but failed. Sending Shonnie to Radiology, R—— made a note to involve Dr. William W——, a general surgeon at the local hospital, pending the results of a mammogram.

Rather than a mammogram, which is an X-ray, Shonnie had an ultrasound scan on April 3. Ultrasound does better than X-rays at resolving the dense breast tissue of young women. It was the radiologist's opinion that the object he saw was a complex cyst. But since Dr. R——'s attempt to aspirate fluid had caused some inflammation and made the cyst a little harder to read, the radiologist called for another scan in a month's time. A biopsy might be in order if the cyst didn't clear up. Telephoning the information to Shonnie, R—— again asked her to get in touch if the cyst got any bigger before the next ultrasound.

Shonnie let nearly three months elapse before she made the appointment for her follow-up scan. No doubt she hoped that the lump was indeed a cyst and would go away. The cases of breast and ovarian cancer in the Medina-Martinez clan should have been a wake-up call to her medical providers long before this, but nothing about her relatives appears in her records yet, not even a denial by the patient that she had a family history. (Shonnie would not have lied if she were asked.) It suggests that the doctors never inquired. A twenty-something-year-old with breast cancer? Very unlikely. When Marianne and Iona are questioned about Shonnie's diagnosis, they bring up the mistake the doctors made by calling it a cyst, and they magnify the time until the doctors corrected their mistake, yet they forget or collapse the time that Shonnie herself let go by, as her worry mounted, maybe as *all* of their anxieties mounted.

So now, in late June, the mass was about eight times as large as it was in February, and on the screen it was more solid-looking than before, with some irregular internal striping. The radiologist (a different radiologist looking at the image) termed the mass suspicious and recommended a biopsy. Shonnie was sent to Dr. W——, the surgeon. He becomes the villain of the piece in her relatives' recollection, the other doctors having been forgotten.

The first meeting with the surgeon, on July 14, went smoothly. W——'s case notes, to quote them exactly, say that Shonnie is a real pleasant, twenty-seven-year-old Hispanic lady. He recorded that the skin of her right breast was retracted (pulled back) at the place where he felt a two-centimeter mass. The biopsy was scheduled for the next day, and he explained the procedure to Shonnie and her husband, including the risks of scarring, bleeding, or infection from the excision. Shonnie wondered if the procedure could be done by laser surgery instead. No, W—— said. The couple agreed to go ahead. He noted that she was a Jehovah's Witness not wanting to receive any blood products, which was not a problem for him.

After the patient was sedated in the operating room, W—— cut out as much of the lump as possible and sent the fragments to the lab to be analyzed. It was an invasive carcinoma, originating in the milk duct and breaking out into the breast tissue: a malignant tumor.

Shonnie, Michael, Joseph, Marianne, and Iona arrived at the doctor's office on July 17, two days after the biopsy. They already knew she had cancer. Shonnie needs a mastectomy, Dr. W—— explained. There were two options, modified radical mastectomy or partial mastectomy, one more severe than the other, but she needed surgery, no question, with additional treatment likely, chemotherapy and/or radiation. The family said they wanted to get a second opinion in Denver. W—— said sure, adding that he would be happy to take care of Shonnie here, in Alamosa, after she and her consultants decided on the best course. W—— asked to be informed of her plans the next week. The meeting lasted twenty minutes. He didn't seem to suspect that she was headed in another direction entirely, that she sought another kind of provider.

If breast-cancer treatments were not toxic and disfiguring, and if they cured every woman, then alternative therapies would not allure a substantial minority of the patients. As it is, one of three Americans reports using alternative or natural medicines, also known as complementary medicines, because patients will add them to the conventional approaches rather than substitute them. Cancer patients are particularly prone to hedge their bets this way, and good oncologists don't object unless there are unfavorable drug interactions. The murkiness of the disease's causes, the imperfect prognoses, even the occasional spontaneous remissions can shake a woman's faith that her oncologist knows what he's talking about. A good doctor, recognizing the uncertainties, moves carefully.

The two parties in this particular cancer dialogue were very far apart when they started their negotiation on July 17. W—— had a lot of experience with cancer, but he was a general surgeon nearing retirement at a small, rural hospital in Colorado and was used to having his way. Shonnie, well, she was Shonnie. She wouldn't do synthetic medicine, said Marianne, who was a strong influence on her daughter's decision. She didn't want medicines that had been tested on animals. She prayed on it. She said, Give me something else. Shonnie's uncle Bill championed the alternative approach from his more New Age perspective. Their influence aside, the decision Shonnie made was relatively easy because she could not, would not, have her body marred. As she told her sister, If I get to live, if I have the remission, I want to look decent.

The person Shonnie went to see in Denver was a homeopathic practitioner named Larry K——. Homeopaths try to stimulate the so-called healing energy fields of people whose energy fields have been disrupted by disease. The theory is that if you are able to mimic the symptoms of a disease, at a very mild level, by administering tinctures of natural medicine at very low doses, the body's vital force will respond. It's immunotherapy without good biology behind it. Popular in Europe and America during the nineteenth century, homeopathy held that like could cure like. The rival school of medicine was allopathy, which tried to cure disease by cruder means, unrelated to the style of the symptoms. To the practicing homeopath, the scientific medicine that came to dominate the twentieth century is allopathy, and cancer chemotherapy is quintessentially allopathic. Doctors with conventional medical training think that homeopaths are no better than quacks, and that their tiny bottles of diluted herbal medicines are placebos.

It's not clear how the Medinas found Larry K——. He was said to have cured a cancer patient they knew, or knew about. At his office in Denver he treated his clients—he did not call them patients—while his wife was in charge of ordering and preparing the special pharmaceuticals. K——, who is still active, travels beneath the radar of Internet search engines. According to a state government website, he was disciplined by Colorado for practicing acupuncture and chiropractic medicine without a license, the only official sighting of him.

Ten years after treating her, K—— said he couldn't recall much about Shonnie Medina. She had a viable cancer, moving quickly, he said. It was deadly, the way we saw it. It wasn't an average breast cancer. Most of these we can turn around. . . . K——'s voice drifted off for a moment. She didn't feel it was that disastrous. We can wait, seemed to be her attitude.

How did K—— know her cancer was growing quickly? We do measurements, he said, referring to insights from homeopathy and Chinese acupuncture. Cancers have predestined pathways. Lymph points, for example. We have measures of toxicity in blood cells, the red blood cells, if it's getting worse.

What kind of measures?

Traditional medicine can give you names, but we don't need to know them, because we're not allowed to diagnose. His words had a sleepy, veiled quality, as if he might be dreaming a conversation. But I can tell if a degenerative meridian is changing. The lymph points for cancer have a rate of degeneration. For diseases like cancer and arthritis, I measure all the meridians.

Didn't your measurements show that Shonnie was getting better because of your treatment? At least that's what she said.

That's not my recollection, K—— said sharply, as if slapped awake. I confirmed what the traditional doctors said. . . . We needed a sense of urgency. . . . His voice was veiled again. He sounded like a clogged drain. [Later] I sent her for natural health [treatments] to people in Mexico, to get her some opinions, some more education. She was exploring it from a natural standpoint. . . .

Here are some of the alternative medications that Shonnie took for her illness during the summer and fall of 1997, followed by Larry K——'s short explanation of each:

- Bioactive Botanical Complex. For muscle repair, sore muscles.
- Radiation (a liquid). To restore the body after exposure to toxic X-rays, ultraviolet rays, rays from TV sets, etc.
- Serotonin. For stress and sleep disorders.
- Lymph Tonic III. To help the lymph glands eliminate swelling and toxicity.
- Candida Albicans 200X. An antifungal agent. In diluted form it mimics the effect of Candida infection and stimulates the immune system.
- Sciatica-HP. To reduce swelling. She was a skier, said K———. It appeals to certain parts of the body, to the electrical values. It tells the nerves to settle down.
- Botanical Gingerplex II. It builds energy when the adrenals are worn down.
- Shark Cartilage 2X. An anti-tumor agent, to dampen the growth of her tumor's blood vessels.

Meanwhile Dr. W———'s nurse in Alamosa attempted to find out what was going on with Shonnie. On November 7, about four months since their last visit, the family reappeared in his office. The gathering was the same as before, minus Joseph Medina, according to W———'s notes, although Iona recalls that Joseph was there because he scolded her afterward for her part in the encounter. Either way, it was the women who did all of the talking to the doctor.

W——— dictated an extensive, self-protective account of the stormy meeting. He appears to want to be covered in case he were sued. They tell me, W——— began, they have been having homeopathic treatment by a Dr. C——— [*sic*] in Denver. They brought some papers that showed some type of medication-type of test that showed regression of her "disease." They wanted to discuss all of this at length concerning whether or not I could do a test to see if her cancer was gone or if it was better. She told me she still had a lump on her breast.

Shonnie told the doctor of a machine that indicated her cancer was going away. The machine was probably K———'s meridian health monitor, an electrical device. The way K—— described it, electrodes are placed on the patient's body to capture the flow of energy along the meridians of the arms and legs. Again, this relates to Chinese acupuncture. The patient squeezes the handle of the device while the examiner passes a sensor over the meridian lines. A computer analyzes the results for excesses or deficits of energy.

All of them were quite surprised and at a loss, Dr. W—— went on, to understand why there was not some type of test that we could do to know whether or not she still had cancer. The doctor explained that imaging tests like MRIs and X-rays can show if a mass is present or not, but that only a biopsy could tell if the mass was cancerous.

Addressing Shonnie with a lot more force than before, the surgeon reiterated that she should have a mastectomy. He again went over the two surgical options. If the tumor were left untreated, he said, this cancer might possibly erode through her breast into the skin and out onto the skin and cause pain, bleeding, stinking, and it would be a very terrible way to die. Surgery was the only justifiable and scientifically proven alternative. Chemotherapy and radiation might be needed too.

Marianne thinks this was when he lost Shonnie altogether. W——'s harsh, negative message, she said. He looked at her like she was a mature adult. I wish he'd been more patient with her. You don't hand a ball of fire to a patient like that. It was like he had Shonnie cornered. You just can't demand it, if you want to get through to her. But getting through to Shonnie might have been impossible. Told what she must do, she was the *converso* rankling before the Old Christian, the Indian resisting the *española*, the Hispano resentful of the American, the Witness rejecting the world's authority while living within the world.

Marianne also recollects that he used the word quack. W——'s notes say he took the high road—that he asked for proof that K——'s treatments were effective. I explained to them I did not concur with making therapeutic decisions based upon anecdotal cases. I told them I was not an expert with alternative medicine and that I was not skeptical of it [*sic*], but I certainly was willing to consider any data they might be able to provide or that an alternative medicine provider might be able to provide to show that his or her treatment was in fact valid.

Both sides dug in their heels. Shonnie said that she would prefer to die intact than to appear dead while she was still living. W—— said that was a common reaction but that most women did very well emotionally afterward adjusting themselves with either partial or complete mastectomies. His case notes make no mention of reconstructive surgery, but Marianne recalls it was mentioned. Shonnie told me, the doctor continued, that she felt that taking part of herself off would destroy her spirit.

Iona spoke up. You can't guarantee that your way will save her. Can you say it? Say it. This part isn't in W——'s notes.

The matter of the surgical costs must have come up. All of a sudden, Shonnie got to her feet. Holding a batch of forms from the hospital, she threw them on the floor. Her face was red; the veins were popping out on her face. She was madder than her mother had ever seen her. She started to walk out the door, and she turned and shouted, You just want my money! The people in the hallway looked up in alarm. Iona chimed in, To you she's just a walking dollar sign!

I was angry by this [*sic*] and quite openly so, W—— countered. I told her that I thought that was an incredibly insulting remark to make to me and that I did not think it was reasonable or fair at all, telling all of them that I think they did not know me at all enough to make that kind of judgement. She later apologized for making that remark, prior to ending our discussion. . . .

She stated on several occasions that she knew it was not my responsibility for her choice, said W——. I explained to her that I certainly felt responsible anyway but I certainly understood her choice and told her she certainly was within her right to make her own choice as long as it was reasonable and a well-supported one. That was where I could not agree with her.

If the doctor had been more conciliatory, he might have tried to explain to Shonnie the difference between a scientific study of a cancer treatment and the anecdotal reports from alternative practitioners. Examples of a treatment's success are like the numerator of a fraction.

They are a good start, but the only way to validate them is to look at the denominator, which includes the treatments that may have failed. To do that, you conduct an experimental trial. You know the number of patients going in, what each person was given, and you measure the outcomes. That's science. On the other hand, when you hear of cures from complementary or alternative medicine, the universe of patients is not defined and the failures are not counted.

Dr. W—— wrapped up his statement. I invited them to let me know if there was anything further I could do and also asked them to please keep in touch.

Shonnie stalked out to her car. She sat in the vehicle by herself for a few minutes. The others knew to leave her alone. After she got home, she went straight to her bedroom and shut the door.

A week after the meeting with W——, Dr. R—— came back into the picture. She tried to placate Shonnie over the telephone. The cancer was aggressive, R—— emphasized. Shonnie said she would rather die than be maimed. R—— finally persuaded the stiff-necked young woman to consult an oncologist in Denver. To see what he might have to offer, said R——.

Before going, Shonnie had an ultrasound scan and a mammogram, which showed a multilobed tumor in her right breast, plus a new and suspicious mass in her left breast. Bilateral disease is one of the hallmarks of BRCA carriers, just as her youth was a clue. Three years after the discovery of 185delAG and other BRCA mutations, the probability that Shonnie was a carrier of a mutation was not discussed. Of course, the case by now had moved far beyond genetics.

Dr. S——, the oncologist in Denver, was very thorough when he met with Shonnie in early December 1997. Among other things, he recorded that four of her aunts had died of breast cancer. He reviewed all of her tests and scans and performed another ultrasound. But the oncologist failed to make headway with Shonnie and her husband when he urged not only a modified radical mastectomy of her right breast but also a biopsy and probable mastectomy of the left. Worriedly, Dr. S—— wrote to Dr. R——, I do hope she gets some follow-up someplace and gets these cancers taken care of pretty quickly since they are showing fairly aggressive behavior.

Shonnie abandoned the San Luis Valley medical system. She didn't answer an entreating letter from Dr. W—— that acknowledged her frustrations and reaffirmed his position about her care. Shonnie followed Larry K——'s advice instead. In late 1997 and 1998, like an exile questing for home, scorned and persisting, getting tired, getting sicker, Shonnie traveled five times to Tijuana, Mexico, for treatment, to a place called the Bio Medical Center.

The Bio Medical Center offered the Hoxsey Herbal Treatment, which was the most famous, or infamous, alternative cancer remedy of the twentieth century. Invented in the 1930s by Harry Hoxsey, an ex–coal miner, it consists of a tonic that is taken internally and an external paste or salve, both derived from natural compounds. Hoxsey's formula was secret for many years, which only increased its appeal. Hoxsey warred constantly with medical authorities in Texas and other states, and he saw his last U.S. clinic shut down in the 1960s, prompting his nurse and lead acolyte, Mildred Nelson, to move the Hoxsey therapy across the border, where renegade cancer clinics operated freely.

Shonnie and her mother read up on the Hoxsey therapy. Impressed by the Bio Medical Center's beautiful building and welcoming staff, Shonnie undertook her first treatment in Tijuana in December 1997. Each course lasted two to three weeks. Arriving as an outpatient each day, she would be given the special tonic, or receive vitamins intravenously, or have wraps containing the caustic salve applied to the skin of her chest. Shonnie also took tamoxifen pills, an established drug for breast cancer having few side effects, though usually tamoxifen is given in combination with chemotherapy. At night she'd stay at a motel in nearby California or at the homes of supportive Jehovah's Witnesses. The family laid out thousands of dollars during this period. Shonnie had insurance through her husband to cover her care in Alamosa, but the unauthorized treatments by Larry K—— and the Bio Medical Center and the special medications were not covered. The Komen organization gave the Medinas a little money, as noted earlier, and a well-to-do Witness provided them with a car and gas.

In the spring Shonnie became pregnant, much to her surprise and distress, because she and Michael had always used birth control. She visited an ob-gyn specialist named Dr. Christine S—— in Pueblo, outside of San Luis Valley. Shonnie had already been advised that if she were going to carry a baby, she must have her disease and her medications evaluated. Dr. S—— therefore remembers Shonnie not for her pregnancy but for her refusal to undergo conventional treatment. I added my voice to the others, the doctor recalled, but she had very strong feelings. She did not want to mutilate her body. I felt bad. Such an attractive young lady with a bad breast cancer. The ob-gyn specialist saw Shonnie only that one time. Shonnie miscarried not long afterward.

At home between sessions at the Bio Medical Center she followed a diet based on Hoxsey's prescriptions. Fresh fruits and vegetables were emphasized, in addition to vitamins, laxatives, and antiseptic washes. Shonnie prepared her meals from a mimeographed copy of the approved cookbook, which is stern and sweeping on what not to eat. Prohibited are pork, tomatoes, vinegar, carbonated beverages, alcohol, white sugar, white flour, salt (except for the salt contained in bread and cheese), animal fats, and all canned foods. A sure way for patients to avoid forbidden pork—it could find its way into processed meats, they were advised—would be to buy their cold cuts in a Jewish market. Jews do not use pork, the book pointed out. What was the problem with pork? According to Marianne, it interfered with the Hoxsey tonic.

Fond of spaghetti and pizza, Shonnie hated to give up eating tomato products. When she would have a lapse, say the heck with it and have a pizza, she'd feel guilty the next morning. She'd cry and wonder if it all was worth it. Marianne invented a spaghetti sauce that substituted red peppers for tomatoes, and she fed and comforted her child. I think sometimes she didn't want to live, Marianne said. She'd get in those moods. . . . But other times she was so *positive* about her illness.

In July 1998, Shonnie called and asked the medical staff in Alamosa to refill her tamoxifen prescription. Dr. R—— responded by registered mail, saying that Shonnie would have to come in for an appointment. If she wanted R—— to be her doctor again, surgery and chemotherapy would be recommended. When Shonnie got the letter, she telephoned immediately. You are not my doctor any longer because you did not agree with my treatment plan, she said.

She wasn't getting better. You could tell she was sick because her hair was dry, said Jackie Williams, one of her friends—but she'd never tell you how sick she was. Here is Shonnie glancing into a video camera, wan and a bit puffy. She smiles and moves quickly out of frame. Here are photographs of the family visiting Sea World in San Diego, Shonnie bundled up against the ocean breeze, refusing to act sick. During a break from treatment, they went to the Universal Studios amusement park in Los Angeles. Shonnie loved the violent lurches of the Back to the Future ride, which shook her out of her pain. For by this time the tumors pressing against her skin seemed to scald her from within. It burned, it hurt so much she couldn't touch herself, said Marianne, but on that ride she was driving and having a blast.

Celina Gallegos, a friend of the family from the Denver area, visited in the fall of 1998. I remember wanting to hurry and leave, to not disturb her after finding out she wasn't seeing people, Celina wrote in an e-mail. But she apparently was feeling up to company that day. I remember her wearing a robe and that Iona had done her makeup while we waited. Shonnie was a pleasant surprise. What shocked me I suppose was the full body of hair she had flowing. I knew she hadn't gone through chemo but somehow I still expected her to be, well . . . less Shonnie. She was actually pretty animated in a subdued sort of way, if that makes any sense. I faded into the background while she went back to the childhood memories she had with Rod (my husband), and although I can't identify an exact moment, it's when I realized she was being strong for her family and had come to terms with whatever point it was she was at in her life. . . . [She] was very down to earth and matter-of-fact and if it weren't for her being dressed in a robe and appearing a little tired after the visit; well, she didn't seem to have been dying at all. Shonnie certainly may have been a patient to cancer, but I never saw the side of her that was a victim to it.

Some of those close to her complained they weren't being allowed to see her. Her cousin Shannon, for one. We had to sneak to the house, said

Shannon. Marianne would say, She'll be fine, she's doing good, Shannon's mother, Chavela, added. They kept it private for a long time.

We kept people from seeing her? Marianne was indignant. No, it was *her*. She didn't want them to see her sick. She didn't want visitors unless she could be made up—her hair, her nails. She'd say, They just want to be nosy. Some people brought Shonnie balloons or presents as a way to get in, but at the end we wouldn't take them.

She was picky about visitors, said Iona. The sincere ones, yes. Some pretended that they cared but they were really coming because they wanted to see, What does she look like?

When Shonnie's favorite young cousin, Priscilla, showed up one day, Shonnie let Priscilla come in. But Priscilla was crying so hard afterward that Shonnie said, So! That's why! I don't want that. I don't want anyone else in here.

Her tumors, ulcerating, had broken through the skin, making a noticeable odor. During her last treatment session at the Bio Medical Center in November, when the wraps and salve were changed, Marianne was amazed to see pieces of tumor falling off and other pieces emerging from the tissue in starburst patterns. It looked like a cauliflower with fingers, she said. The cancer was growing so fast it was fungating, or outstripping its blood supply, causing the most greedy cells to die. This also indicated that metastasis was well advanced, the tumors having seeded everywhere in her body. Rather than averting her gaze, Shonnie asked that pictures be taken of her emergent tumors. She thought it was important to document her disease, Marianne said.

One day when we were meeting at the restaurant, Joseph asked his wife if Shonnie would have approved of my publication of her medical records. The photos of her disease had been burned, but a stack of records remained. Sheets of their daughter's body, spread on the table for everyone to look at. Yes, absolutely, Marianne said.

She was terribly vain about her body, of course, but her body was all she had. Remember—there is no soul for a Jehovah's Witness. The spirit cannot be divorced from the body, nor can the spirit lord over the body's corruption, lashing the flesh for the spirit's own satisfaction. The body sickens and rots because of original sin, the almost-genetic inheritance of man's imperfection. Life may be halted, its substance suspended, until the body is restored in the new system to come. The self is going to be restored.

Cancer made Shonnie feel like a reject, like she was ugly (Marianne's words). Outside, beautiful. Inside, ugly. Living a lie. If she had had the surgery, she'd be ugly both inside and out—intolerable. When the imperfection broke through at last, fascinating and awful to view, Shonnie was already dead and preparing to be renewed. As her new body took form, the world that she was leaving was welcome to have the pieces of the old one.

During her last treatment in Tijuana, Shonnie was wracked with coughing, and she developed a fever. A doctor at the Bio Medical Center urged the family to get her into a hospital in California. Shonnie entered Chula Vista Hospital on December 4, 1998. While treating her for pneumonia, the Chula Vista doctors prevailed upon Shonnie to have radiation therapy for her cancer. She tried it but she wasn't happy with it, Marianne said, and she refused chemotherapy.

Shonnie was terminal—nobody was avoiding the prognosis now. It was while she was at Chula Vista that Shonnie had the BRCA test and found out she carried 185delAG. The genetic coloration of her disease led her parents to think that it couldn't have been stopped, regardless of the therapy.

Feeling a bit stronger, Shonnie made visits to the children's ward on the cancer floor. The children wondered how this pale, striking woman in a gown, religious pamphlets under her arm, how could she have cancer and yet have all of her hair? Shonnie would return to her room in tears. Why do kids that young get cancer? she said to her mom. Her cousin Shannon showed up in San Diego unannounced, and the two joked about Shonnie's lack of tan. Girl, get some sun!

Shonnie wanted to go home. Because the pain was constant, first she was transferred to the Penrose Cancer Center in Colorado Springs. Pain medication was set up for her, and it was arranged for her to become a hospice patient on her return to Alamosa. Around Christmastime Shonnie traveled over the jagged rim of the mountains into the San Luis Valley, the Hispano homeland, for the last time.

She cut out all visitors outside her family. She would not have any hospice nurses or volunteers helping her who might already know her. Marianne was with her all the time, and Michael, and Iona whenever she could, along with Iona's husband. Michael put up an exercise bar over the bed, because Shonnie still had an appetite and tried to keep her body in shape. She worried about atrophy, said Marianne. Even near the end, she loved to eat. She refused to get bedsores.

Life goes on, Shonnie would say to Marianne. Her father, her sister, her husband were having a difficult time. They'll be OK, Shonnie reassured her mother. To Iona she said, Everything goes to you, Kiddo, meaning the pieces of property Joseph had painstakingly assembled in Culebra.

Early on January 27, 1999, the hospice nurse thought this would be the day. Shonnie had been getting oxygen because of her chain-coughing, the intractable cough that sometimes occurs in advanced cancer cases. Telephone calls were made. Michael went out to buy some bottled water for her. Iona arrived, and Shonnie said hi to her in a weak voice. Iona went over and braided her hair. Shonnie began to slip in and out. When Michael got back with the water, he threw it against the kitchen wall in despair.

Joseph didn't make it in time.

The others wanted Shonnie to keep fighting. Her mother didn't. Shonnie said, Mom, I can't do this anymore. I have to rest.

She looked at me, and I could see that her eyes were cloudy.

So I told her, Rest. Just rest, Shonnie.

Marianne added that Mildred Nelson, the director of the Bio Medical Center, died the very next day. She'll never forget the coincidence of Mildred Nelson dying. At eighty, from a stroke.

Chapter 9

WHEN HARRY MET STANLEY

O ne September day in 2001, a group of genetic counselors was having lunch in Denver. They were Teresa Castellano, Lisa Mullineaux, Lisen Axell, and Jeffrey Shaw. The four friends, who represented different medical centers in Denver and Colorado Springs, used to meet periodically to talk shop. Today Teresa Castellano mentioned the oddity of having uncovered a Jewish breast-cancer mutation in a rather young Hispanic woman. I said, I have a patient with the 185delAG mutation, Castellano recalled, and she's only in her forties, and she's from the San Luis Valley. Lisa Mullineaux put in that she had seen a couple of cases like that.

Then Jeff Shaw spoke up. Although he had met Marianne Medina in the hospital two years earlier and may even have glimpsed Shonnie, Shonnie's case didn't come to his mind—but Shaw did have another 185delAG carrier of his own to report. Since BRCA testing had become available, each of the genetic counselors had seen one or two Hispano breast-cancer patients who were 185delAG carriers. Going back to their files, they learned that all of the women had roots in the San Luis Valley.

Mullineaux spoke next with Ruth Oratz, a New York University oncologist who was then working in Denver. Those people are Jewish, Oratz told Mullineaux. I'm sure of it. Oratz was a close colleague of Harry Ostrer's. Convinced they were onto something interesting, the four counselors enlisted a genetic scientist, Sharon Graw. Graw's analysis of the mutation confirmed the telltale Jewish spelling, and the group published its finding in 2003 in the journal *Cancer.* Their research paper—the first report of 185delAG in Hispanos and the foundational document for this book—was titled "Identification of Germline 185delAG BRCA1 Mutations in Non-Jewish Americans of Spanish Ancestry from the San Luis Valley, Colorado." Germline means inherited. By non-Jewish, the researchers meant that the six women in the study who carried the mutation claimed that their families were Catholic. They denied being Jewish or having Jewish heritage.

Some of the subjects had started to wonder, though. One woman, after quizzing her relatives, told the researchers that she suspected she had Jewish ancestry after all. Another subject, a patient of Mullineaux's named Beatrice Martinez Wright, went further. With the zeal of an investigative reporter chasing two stories at once, Bea Wright went looking for cancer cases and evidence of Jewishness in her family tree. Everything that this chapter contains is contained in Bea Martinez Wright, either in her cells or in her awakened sense of her Jewish heritage. Shonnie may have left the stage, but the long-running story of the secret Jews of New Mexico is about to intersect the Medina family because Bea Martinez Wright was Shonnie's first cousin, once removed.

Bea's father, Luis Maximo Martinez, who was called Max, and Shonnie's grandmother Dorothy were brother and sister. They were born a year apart in the large brood produced by Luis and Andrellita Martinez. Recall that Andrellita Martinez had breast cancer, survived it, and went on to die of another cause. Almost certainly, the 185delAG mutation was introduced to the cancer-plagued family by way of Andrellita. Almost certainly, she was the obligate carrier, a person to whom a mutation can be traced without proof of a DNA test. It had to be either Andrellita or (less likely) her husband.

Max, Dorothy, and the other offspring of Luis and Andrellita were raised in Culebra in the 1930s and 1940s, but Max Martinez was restless and left for bigger places. Large families outgrew little properties;

more people left during the postwar years than stayed. Max settled first in Colorado Springs and then in Pueblo, Colorado, an obligate but unknowing carrier of the mutation like his mother. Max and his wife, Rosalia—her maiden name also was Martinez—had seven children. Beatrice was born in 1954 on a farm where her father was a migrant laborer. Andrellita traveled from Culebra to the place where Max was working in order to deliver Bea.

Meanwhile, Dorothy's oldest son, Joseph, an obligate carrier too, remained on the land in Culebra. Three years older than Bea, Joseph Medina had little if any knowledge of his first cousin. Bea's own memory of Joseph was better, because she made visits to Culebra to see their grandmother. Bea even remembered that Joseph had a little daughter, Shonnie. Married twice, with three children of her own, Bea resided in the Boulder area, a long way from the San Luis Valley.

Beatrice Martinez Wright was diagnosed with breast cancer in 2000, when she was forty-five years old. Shonnie had died the year before, but Bea didn't know that then. Bea had a mastectomy of her right breast and chemotherapy. She had the good fortune to be treated at a major medical center in Denver. Having heard of cancers among the women on her father's side, she informed her doctor, and considering that, plus her relatively young age, the doctor recommended that she see a genetic counselor, who was Lisa Mullineaux. That is how Bea learned that she carried the BRCA1.185delAG mutation and that she was descended from Jews.

In the fall of 2001, Bea scheduled an operation for removal of her healthy left breast as well as her ovaries, fallopian tubes, and uterus. These were aggressive prophylactic measures for a person of her time and place. Before she would have the surgery, however, she decided to return to the Valley and to New Mexico to warn her relatives, as many as she could find, about the mutation in the family. She also wanted to tell them the exciting news that the Martinezes and Medinas were related to Jews. I was mostly concerned with our cancer history, she said. This was flowing in our veins. I thought, We've got to take care of it because it hurts when you hear of another one gone down.

Bea also hoped to do some genealogical fact-finding on the trip. I have sixty first cousins, some I never knew I had, she said. I made the trek because I needed to know where I was from. She drew up a pedigree of the cancer cases she'd gathered, and she made a stack of photocopies of the chart and other material to hand out to her relatives. About the mutation she wrote: "This is the same mutation found in the Ashkenazi Jewish Ancestry! So somewhere in the family line, from past generations I'm of Ashkenazi Jewish Ancestry!" At that time most geneticists and genetic counselors did not know that 185delAG affected Jews world-wide, not just Ashkenazim, and that the route of transmission to His-panos was probably via Sephardic Jews.

When Lisa Mullineaux told Bea that her mutation was Jewish, Bea recalled a magazine article she'd read a few years earlier. It was about the secret Jews of New Mexico. Bea compared what she remembered of the article with peculiar incidents from her childhood. There were odd goings-on in her grandmother's household. Andrellita and other female relatives would light candles on Friday evenings, as if for a furtive Shab-bat. During a funeral, or was it afterward, during the period of mourn-ing, the women would cover all the mirrors in the house. When they swept the rooms, they'd make a pile of dust in the center of a room, never at the doorways. And there were some other practices that, look-ing back, seemed unusual. Were these things Jewish? As far as Bea knew, her grandmother had been a sincere Catholic. Her father, Max, dis-missed her questions out of hand.

In the library and on the Internet, Bea read a number of accounts of New Mexicans who had rediscovered their Jewish heritage. She learned that the phenomenon was called crypto-Judaism and that it had started with the suppression of Jews in Spain and had carried over to the New World. Although one expert maintained that crypto-Judaism had died out in New Mexico centuries ago, others declared that memories like Bea's proved it was still alive.

By the time she and her husband left for the San Luis Valley, Bea had become a believer not only in the Martinezes' Jewish ancestry but also in a Hispano Jewish identity. These are two different things, and a person with a rock-solid religious foundation might embrace the first, accepting a Jewish forebear or two, without buying into the second. But Bea was a disaffected Catholic, you have to understand, and had been for a long time. The Catholic Church kind of sucked, she said, apologizing for her bluntness. I didn't believe it, I'm not a Catholic, she said. But Bea was too modern-minded to become a Jehovah's Witness like her cousins. Rather, she described herself as a Gnostic Christian who could talk to God anywhere, and in addition she admired the approach of the New Age psychics. Unencumbered by tradition, Bea could claim a Jewish identity for herself, as many Hispanos were doing even without the evidence of DNA. Admittedly it is hard to get a fix on Bea's beliefs because her mind was swirling. Her cancer scare and the 185delAG mutation had upended her understanding of her family and its history, that much was certain. She drove off to tell her relatives.

And while she is en route, it is appropriate to introduce New Mexico historian Stanley M. Hordes, the scholar most responsible for the rediscovery of Jewish heritage by Hispanos. Though Bea would not have recognized his name, the information she had gathered about crypto-Judaism was generated in large part by his efforts. If Harry Ostrer stood for nature, i.e., the biology of Jewishness expressed through DNA, Stan Hordes represented its counterpart, nurture, or Jewish identity expressed through the cultural landscape of northern New Mexico and the San Luis Valley. If Harry was a fox, always seeking new insights and new collaborators, Stanley was a hedgehog, never wavering from a single idea. Working independently, the two researchers converged in Bea Wright, because she was one of a handful of Hispanos who carried both a Jewish gene and what Hordes called a Jewish consciousness.

Stanley Hordes was the official state historian of New Mexico during the early 1980s. His job in those days was to manage the archives in Santa Fe and browse around in New Mexico's colorful past. His name (pronounced HOR-dees) sounded Latin, as if he might be from around here, and he spoke very good Spanish, but in fact Hordes was an outsider to the academic establishment of New Mexico, an Ashkenazi

Jew from back East. He was an independent thinker, a dogged worker, something of a loner, and sensitive to slights. Mrs. Hordes didn't raise a thick-skinned child, he once said about himself.

Part of his job for the state was assisting people with their genealogies. I began to receive some very unusual visits in my office, Stan recalled. People would drop by and tell me, in whispers, that so-and-so doesn't eat pork, or that so-and-so circumcises his children. Hordes, who had written his doctoral dissertation on the crypto-Jewish community of seventeenth-century New Spain, pricked up his ears. The hushed rumors he heard in the state archives were echoes of accusations that had been made in the offices of the Spanish Inquisition in Mexico City in the 1640s. To accuse a high-ranking citizen of Judaizing was a useful political tactic and might well strike the mark. For the Spanish settlers of the Americas had included genuine Catholic converts, aka *conversos* or New Christians and their descendants, but also false converts, the crypto-Jews or Judaizers, who maintained their faith in camera. Crypto-Jews existed among the nobility and political leadership in Mexico City. Although the Inquisition on two occasions campaigned to root out the Judaizers, it is clear from the records of the trials that their practices endured, even in the face of executions.

According to Hordes's research, the crackdowns probably caused crypto-Jews to venture farther up the Rio Grande to frontier outposts where they might avoid persecution. The period of the colonizing of New Mexico coincided with such a crackdown. Juan de Oñate, New Mexico's conquistador-founder—a man who whipped himself as he crossed the Rio Grande out of respect for the Easter season—came from a *converso* family, as did a number of members of his expedition. These periods were exceptional. On the whole, the authorities in Mexico City were tolerant of *conversos*, and in Santa Fe they were even more so. All pressure on crypto-Jews ceased after 1700, when the colony entered its second phase, following the Pueblo Revolt, and Judaizing was a forgotten issue when New Mexico passed into American hands in the 1840s. Until he began hearing the rumors to the contrary, Hordes assumed that the secret Jews of New Mexico had withered away, history having made no mention of them for centuries.

But when he went into the field in pursuit of the stories, Hordes found evidence to change his mind. A watered-down variant of Judaism appeared to have survived well into the twentieth century, its trail still warm. "Children growing up in the 1940s, 1950s, and 1960s," Hordes wrote, "witnessed their parents lighting candles on Friday night, refraining from eating pork, or slaughtering their meat with special care not to consume the blood, but were not told the reason for these observances." Some of the Hispano families, such as Bea Wright's, seemed to have forgotten the Jewish significance of the customs, holding onto them only by habit, while others knew full well what they were about; or at least the older generation had known, if their children were to be believed. "It was only when their suspicions were aroused decades later," Hordes wrote, "that they asked their elders, who reluctantly answered, *'Eramos judíos'* (We were Jews)."

Stanley gathered many such testimonies. His informants took him to backcountry cemeteries and showed him gravestones with six-pointed stars. They brought out objects from their closets that resembled *mezuzah* cases (for holding a sacred Jewish scroll), dreidels (a four-sided toy or top), and other vaguely Jewish items. Hordes became a clearinghouse for stories and signs of crypto-Judaism. He joined the faculty of the University of New Mexico, where his writings and lectures attracted sociologists, ethnographers, and folklorists to the new field. After journalists got wind of it, the coverage jogged more memories and prompted additional believers to come out of the woodwork. Hordes appeared in a National Public Radio documentary, "The Hidden Jews of New Mexico," which had multiple airings and is still requested today.

In the early 1990s, crypto-Judaism was a lively cottage industry worked by an enthusiastic cadre of amateurs and professionals. The Society for Crypto-Judaic Studies put out a quarterly newsletter, and each year it held a meeting in a different southwestern city. The speakers presented historical and literary papers; genealogies and personal revelations were explored. Although Hordes, the society's cofounder, wasn't the only investigator of secret Jews, past and present, he was the most energetic in the field. The challenge, he explained to audiences, was figuring out how to document the lives of people who tried not to leave anything behind.

The old-guard Hispano historians in Santa Fe and Albuquerque were dubious about crypto-Judaism because Stanley Hordes, in the words of one, wasn't of the culture, and because the evidence he offered was anecdotal and ambiguous. It wasn't supported, said Adrian Bustamante, a leading New Mexico researcher. It wasn't documented. Even if it's true, Bustamante complained to Hordes, many ethnic groups have contributed to the Hispano people. Why are you focusing on this one?

Why do I do it? Stan would respond, acknowledging a certain hostility to his work. To strip the veneer off. I'm interested in finding out who we are as a community. Strip the veneer off what exactly? The stereotype that all Jews are from Poland and Russia, he replied. For most people, Jewish and Spanish are antithetical. The fabric of Jewish heritage in New Mexico is richer than we thought.

Another skeptic was Fray Angélico Chavez, a Franciscan priest and a venerated figure in New Mexico scholarship. Chavez's book on the origins of New Mexico's families had triggered the popular interest in genealogy in the state, as descendants sought to link themselves to the original Spanish colonists using his charts and his inventory of colonial surnames. Chavez's best-known book, *My Penitente Land: Reflections on Spanish New Mexico*, in which the author imagines his people gestating, laboring, and finally being born when Juan de Oñate and company traverse the Rio Grande, plays up New Mexico's sorrowful Catholic heritage but is silent about its submerged Jewishness.

When Hordes and his revisionist colleagues reread Fray Angélico, however, they noticed a Jewish strain of another sort in the priest's historical narrative. Repeatedly Chavez compares the New Mexico settlers to the high-country herdsmen of ancient Canaan, the Rio Grande River to the River Jordan, and the desert north of El Paso to the Sinai or the Negev. He depicts a second Palestine, the piñons and junipers on the hills taking the place of gray-green olive groves. Here a different band of shepherds ekes out a living underneath cold, starry skies and a fierce God. Chavez even detects a racial symmetry between the colonists, with their black and sometimes frizzy hair, and the Semitic peoples of the Middle East who had migrated to Iberia and thence to the New World.

Previously this had been taken as a romantic, Judeo-Christian conceit, but now it was asked, What's with all the Jewish allusions? Did Chavez mean to signal the role of Jews in the foundation of New Mexico? The priest was adamant that he had no hidden agenda. Crusty in his old age, Fray Angélico dismissed the importance of crypto-Judaism. He merely meant to link the settlers of New Mexico, Castile, and Palestine, three pastoral peoples on the same latitude, though living at different times. Fray Angélico's disavowal did not completely persuade Hordes and others. They pointed out that his book on the first families of New Mexico included two of his own ancestors who were accused of Judaizing.

All in all, the construction of crypto-Judaism was expanding robustly in the 1990s, until the arrival in New Mexico of Judith Neulander, a one-woman wrecking crew. Neulander, an ethnographer and PhD student, had been eager to adopt Hordes's methods. He duly took her into the field as he interviewed people, and he showed her the gravestones and curious totems of his research, but something went wrong and Neulander decided that crypto-Judaism was a sham. If it had ever existed, it no longer existed—it was an imagined community, she wrote in a broadside in 1994. For the rest of the decade and into the next, Neulander churned out papers attacking Hordes's findings. She accused him of asking leading questions and planting suggestions of Jewish identity in his informants. The reports of Jewish customs were FOAFtales, she declared, an acronym meaning that they came from a friend-of-a-friend. These FOAFtales couldn't be verified, the actual practitioners having died. About the hexagrams carved on the gravestones? A lot of cultures employed that sign. The dreidels? A lot of cultures had fabricated toys like that. Secrecy was a fig leaf over a dearth of evidence. Neulander wrote that, absent a formal Judaic tradition among Hispanos, there were better explanations for the rites and objects being uncovered—vestiges of Seventh-Day Adventism or Pentecostalism, for example, which missionaries brought to the area in the early twentieth century. And since race often lurks below people's opinions about other people, Neulander asserted that the darker-skinned Hispanos were trying to elevate their status by associating with light-skinned European Jews.

Stanley Hordes was stung. Well, just because there are some people who are wannabes doesn't mean everybody is a wannabe, he said. He was badly stung. On her own, the implacable Neulander had disrupted the study of the *converso* legacy in the Southwest, throwing the field into confusion. An *Atlantic Monthly* article in 2000, headlined "Mistaken Identity?," took Neulander's side. Thus, in 2001, while Beatrice Martinez Wright was digesting the news of her Jewish heritage and zealously driving to San Luis Valley to tell her relatives, the scholars of crypto-Judaism were regrouping. The academics had split into two schools and were proceeding with much more care than before.

On the one hand were the sociologists, anthropologists, and field interviewers, who still collected testimony from Hispano households about their possibly Jewish customs, but now the researchers maintained that the truth of the testimonies didn't matter. The historical validity of crypto-Judaism was beside the point. What counted was the individual's effort to formulate his or her identity—a fluid, necessarily subjective process—and the job of the social scientist was to document the person's effort, not judge it. In 2009, Professor Seth Kunin, an anthropologist, rabbi, and leader of the relativist wing, published *Juggling Identities: Identity and Authenticity Among the Crypto-Jews.* "I view culture as undergoing a continuous process of negotiation at all levels and thus being in a continuous state of construction or re-creation," Kunin wrote. By avoiding firm claims and making a virtue of FOAFtales, the relativists gave Neulander little to attack, even as they defended Hordes and disputed her argumentation point by point.

In the other camp, working pretty much alone, was Stan Hordes. He believed there was something essential and immutable in the Jewish identities being rediscovered, even if he wasn't able to define it. But rather than fight with Neulander, Stan retreated to the specialty he knew best, archival and genealogical research. His new tack, when people came to him with their provocative memories and numinous artifacts, was to ask them about their family trees. If Hordes could demonstrate that they were descended from a *converso* or from an accused Judaizer in Mexico or Spain, then it was more likely that a Jewish consciousness had passed down through their families.

The research he conducted in the urban archives of the former Spanish Empire was hot, slow, and tedious, just the place for a hedgehog. Hordes dug backward until hitting a name mentioned by the Inquisition and then tunneled sideways into the marriage and municipal records to see if the names led to other *conversos* or to out-and-out Sephardic Jews. A fruitful method, which he relies on to this day, it helped put a foundation under contemporary crypto-Jews by unearthing their ties of blood to people of the past, while it finessed the problem of continuity of memory. In effect, Hordes was following the tradition that any person with Jewish forebears was entitled to call himself a Jew.

Now for Stanley to move from blood ties to DNA, from genealogies to genetics, was the obvious next step. Culture might turn itself inside out within the space of two centuries—what better example than Shonnie Medina's family history—while a family's DNA barely budges. Genetic analysis can leap over holes in the historical record or forks in the cultural record because the DNA is as true today as it was then. So Stan Hordes ventured into genetics, recognizing how much he didn't know.

In the 1990s, he was approached by a medical student named Kristine Bordenave. She had become interested in pemphigus vulgaris, a very rare skin condition, a type of autoimmune disorder. It happens to be less rare in Ashkenazi Jews than in other ethnic groups. Bordenave told Hordes that dermatologists were finding the disease in New Mexico: a case here, a case there, but they were adding up. A small number of patients was tested, and in the majority the immunological markers of their disease were the same as reported in Jews, pointing to a genetic relationship and shared ancestry. Moreover, two of the patients qualified as crypto-Jews according to Hordes's criteria. They had active Jewish memories, as he put it, and had *converso* forebears besides.

Next Hordes learned that the mutation for Bloom syndrome, carried by one in a hundred Ashkenazim, had turned up in five Hispanic families in the Americas. Still unsure of the significance of genetics, he covered Bloom syndrome and pemphigus vulgaris at the end of his 2005 book, *To the End of the Earth: A History of the Crypto-Jews of New Mexico*. In a footnote about breast and ovarian cancer, he mentioned that 185delAG had been found in San Luis Valley, per the 2003 journal article by the four genetic counselors in Colorado. Lastly, Hordes referred to Y-chromosome analysis, which recently had become popular with ancestry-seeking New Mexicans. Some men's Y chromosomes contained hints of Jewishness.

Promoted online by consumer gene-testing companies, the Y-chromosome probe was a bit of a gimmick because only one slender line of ancestry was accessed, leaving 95 percent of a male's heritage up in the air. Nevertheless, a Catholic priest in Albuquerque, Father Bill Sanchez, had taken the test and by its result had proclaimed himself genetically Jewish. Not only Sanchez but dozens of Hispanos were reported to be descended from the priestly class of Jews, the so-called cohanim. These men carried a stretch of DNA known as the Cohan Modal Haplotype, suggesting that they were related to Aaron, the brother of Moses, which was about as high up in the Hebrew hierarchy as you could get. Father Sanchez emerged as New Mexico's most colorful crypto-Jew. He sprinkled Jewish and also Pueblo Indian rites into the Catholic services he held at his church.

Judith Neulander, now on the faculty of the Judaic Studies Program at Case Western Reserve University in Ohio, watched these developments unhappily. Two could play the scientistic game. Neulander joined forces with a team of geneticists, and they produced a study in 2006 showing that the Y chromosomes of a random sample of Hispanos were apparently no different from the Y chromosomes of contemporary Spaniards. If a significant portion of crypto-Jews were among the Spanish Americans, the DNA of the group as a whole ought to be different, or so she claimed. Hispanos ought to be more recognizably Jewish, but her study showed they weren't.

Turning to Hordes's forays into genetics, Neulander denounced them as pseudoscience and folk taxonomy. "In this way," she commented, "academic use of the label 'Jewish' to determine who is at risk of heritable diseases paved the way for popular use of heritable diseases to determine who is a Jew." Neulander found a study in the medical literature indicating that the DNA markers of pemphigus vulgaris, Hordes's primary example, occurred in people of Mediterranean background and were not limited to Jewish carriers specifically. In other words, the disorder could have come into New Mexico via non-Jewish Europeans.

It's important to understand her criticism about specificity. Neulander had hit upon a weakness in all the genetic research linking Hispanos to lost or hiding Sephardic Jews. It does a scientist little good to track a genetic disorder or a DNA variant from New Mexico back to medieval Spain, thence to make a case for the carriers' Sephardic ancestry. For Iberia had long been a melting pot, breaking down the DNA of racial and ethnic identities. In that relatively open society, an amalgam of Muslim, Jewish, and Christian stock had simmered too long for the strands of DNA to maintain their ethnic consistency. If a genetic detective could not circumvent the admixed Spaniards and follow the DNA farther back in time and space, he was going to be stuck, and his speculations about people's religious origins would be fuzzy.

The one known DNA variant, so far the exception, to pass through the Iberian Peninsula, crossing the Mediterranean and planting its flag in Jewish Palestine, is BRCA1.185delAG. But Stanley Hordes didn't know that then.

Harry Ostrer read here and there about crypto-Judaism in New Mexico. I thought Hordes might be onto something, Harry recalled. More closely, Ostrer followed the scientific work on Jewish ancestral markers in the Y chromosome and mitochondrial DNA (the latter inherited through the mother's line). Though short of a complete picture, the studies reaffirmed that the Jewish people originated in the Middle East—that the Jews, in Harry's words, were who they said they were, regardless of where they might live now.

To show that Jews had partly Middle Eastern DNA did not mean that everyone carrying that DNA was Jewish. Here was the specificity problem again. Just as the medieval Iberians were admixed, so the Semitic peoples of the Levant had exchanged genes before some had migrated west to Hispania, aka Sepharad.

For that reason Harry was increasingly uneasy about the Cohan Modal Haplotype, the block of six markers in the Y chromosome that supposedly represented a straight shot to the biblical Aaron. Many males in the world who weren't Jewish and who didn't want to be Jewish turned out to carry the haplotype. Arabs carried it too. Also, the cohanim stretch of DNA seemed on closer look to have originated before the Judean tribes had, meaning it had spread in the region before Aaron and his ilk ever existed.

But still, wasn't the test useful to a hypothetical crypto-Jew in New Mexico? Couldn't he say, OK, so I'm not for sure descended from Aaron. I still have a Semitic Y, correct?

Yeah, Harry replied, but being descended from Abdul the camel jockey is a comedown from being descended from Aaron the priest.

Located on the far reaches of the Hispano homeland, and having no academic centers to speak of, the San Luis Valley was not a place you would expect to find a flare-up of Jewish consciousness. The Catholic priest in Culebra, Father Pat Valdez, was very skeptical. We are fourteenth-generation here, he said, referring to himself and his siblings. I never encountered Jewish practices growing up. I never heard about this until I read articles about it ten or fifteen years ago. He [Hordes] is reading too much into it, I think.

Father Pat was not ignorant of Jews. There had always been a couple of Jewish families around, he said. He had known a few and could tell others by their names, presumably names ending in berg or stein. One of the factors complicating the role of crypto-Judaism in New Mexico was the much more pronounced historical role of Ashkenazi Jews. During the second part of the nineteenth century, German-Jewish peddlers and merchants had scattered about the Southwest. Some of the Ashkenazi males had married Hispano women; later arrivals procured their Jewish brides from back East or from Europe. At the turn of the twentieth century, Albuquerque, Santa Fe, and Las Vegas, New Mexico, had dozens of Jewish-owned businesses. Thus, when the New Mexico Jewish Historical Society was established, in 1983, it polished up the records of the Ashkenazi immigrants first, and after some hesitation, it got behind the new Hispano-Jewish narrative that Stanley Hordes had introduced. Influenced by Neulander, the leaders of local synagogues were more suspicious. Some of the Ashkenazi rabbis demanded that Hispano would-be Jews go through a formal conversion before they could join.

But again, Culebra was on the outskirts of the controversy. People in Culebra knew about Jews mainly in the sense that Jews were good at business. An absentminded and dare one say innocuous stereotype prevailed. For example, Ricardo Velásquez's father was called Jewish because he made money at his business. Or businesses. My father was a jack-of-all-trades, said the younger Velásquez, a doctor, with a smile. Ricardo remembered that when his father would have lunch with Orlando Mondragon, who ran the garage and repair shop in San Luis, the two businessmen would maneuver to stick each other with the check.

What is amusing about the story of Velásquez and Mondragon is that Orlando Mondragon later came out as a crypto-Jew. This happened after he met Stan Hordes in the early 1990s. Mondragon showed Hordes a strange, six-petaled flower on a headstone in the cemetery, and he described his grandmother's rituals with candles. But even before then, Mondragon said, he had wondered why the unusual given names of men in his family lineage could be found in the Old Testament. His own middle name was Benjamin, not very Spanish. Mondragon's wife had a forebear named Solomon Martinez. Solomon? Were those names just a coincidence? I know a little Hebrew, Mondragon said, and I've been to a synagogue, and I have a love of the Jewish people. He also took a shot at Ricardo Velásquez's father, who wouldn't pick up a check, calling him a shyster. And Father Pat, the priest, he never liked me, said Mondragon.

Still, when you asked around, this elderly fellow Mondragon was the only person in Culebra who could be cast as a crypto-Jew, and even he had by then moved back to New Mexico. Nobody else seemed to know anything about it, or if they did, what they knew was extremely cloudy. That's where matters stood when Beatrice Martinez Wright arrived, charged up about her Jewish heritage and the dangerous genetic mutation confirming it.

Well, before finally hearing about Bea, there's an additional incident to relate. In the mid-1990s, the small-scale cattle ranchers of the southern part of the Valley hatched a novel idea for selling their beef. They'd raise it organically, butcher it according to kosher standards, and market it to Orthodox Jews back East. An agricultural delegation that was visiting from Israel told them it was a great idea. The ranchers, who were both Hispanos and Anglos, formed a co-op, each of them putting up some money. They contacted an Orthodox rabbi, Mayer Kurcfeld, of the Star K Kosher Certification Agency in Baltimore, Maryland. The rabbi agreed to instruct the ranchers in *kashrut*, the Jewish dietary laws, such as applied to the preparation of meat. (Rabbi Josef Ekstein had specialized in this aspect of *kashrut* before taking up genetics.)

Rabbi Kurcfeld came to the Valley in March 1996. He asked the ranchers to bring their Bibles to their meeting with him, so that he might point out the scriptural basis for the kosher slaughtering rites. About sixty ranchers attended. The rabbi showed the type of sharp knife that must be used to kill the animal. In one motion the arteries, esophagus, and trachea are severed, he said. The animal may not be stunned or knocked out beforehand. He demonstrated the decisive thrust. Then all blood must be drained and purged from the meat. And other strict measures had to be followed before the beef could be declared *glatt*, free of imperfection. The ranchers nodded. When their slaughterhouse was ready, rabbis from Denver would be sent to monitor the operation. For a fee.

That evening Rabbi Kurcfeld went to dine at the home of Demetrio Valdez, one of the organizers of the ranchers' co-op. The low-key Valdez and his wife, Olive, live in Antonito, about forty miles west of Culebra. The rabbi remarked, when the subject of washing the dishes came up, that Orthodox Jews referred to it as purifying the dishes. What do you know? The Valdez family had the same saying, a Spanish phrase meaning disinfecting the dishes.

Demetrio Valdez began to open up, for the rabbi's lecture had stirred memories about his grandfather and the butchering of sheep and hogs. We used the whole animal, and we didn't waste the blood, Valdez said. The blood was collected and cooked. Though I think that's from the Native American side. And the knife didn't have to be sharp, and maybe he didn't necessarily use a single stroke.

But the respect for the animal, that was the main thing, Valdez said. And the food was kept as pure and clean as possible. It seemed my grandfather went overboard with cleanliness. Little things had started falling into place for Demetrio after the rabbi's visit. Nobody would admit there was any Jewish in them, he said. Or Native American. It wasn't out in the open.

Off the beaten path of crypto-Judaic studies, this sighting of the entity (not yet an identity) on the Valdez ranch might well be the genuine article, and provide further support that Stanley Hordes was right. But Stan cautioned in an e-mail, It is key to know if they learned about the crypto-Jewish connection internally, or, on the other hand, whether someone from the outside told them about it. It was impossible to resolve that question ten years afterward, since Valdez had learned more about the New Mexico secret Jews in the interim and probably had filled in the blanks of his own story. *Something* buried in the Demetrio Valdez clan had surfaced, and was changing color in the sun. Before you make up your mind about it, remember that indigenous religious practices were suppressed three times in New Mexico's history: the Pueblo Indian rites by the Spaniards, the *penitente* rites by the Americans, and, to some very murky extent, crypto-Jewish ritual by Catholic authorities. What happens to beliefs when they are pushed underground? The weight of yearning in people's hearts may have forged a rough diamond from all three traditions, a syncretic creature that was no longer religious and no longer recognizable.

By the way, the kosher slaughterhouse in the San Luis Valley failed, in spite of a state grant supporting the project. The Orthodox authorities back East became dissatisfied with the processing of the beef. The ranchers struggled to maintain quality control. Essentially the meat was deemed not Jewish enough. The cooperative was losing money when a fire broke out at the facility in 1997, ending the experiment.

The issue of who is a Jew is the hottest one in Judaism, Harry Ostrer remarked. The Orthodox have control of the marriage-license bureaus in Israel and decide who can get married. The Reform Jews have thumbed their noses at the Orthodox, stating that the children of Jewish fathers are Jewish. (Traditionally, only the mother's line counted.) The development of DNA markers for Jewishness by Harry and other scientists might mend the fractures in world Jewry, or it might open up a host of new ones.

His lab at New York University School of Medicine constantly turned aside requests from people wanting to be tested for Jewish ancestry. Harry's work was medical genetics and population genetics, the serious stuff. But he did not fault the drive to know. People are compensating for a century of assimilation, he said, and they're using all the tools of genealogy. Jews have thrived for centuries under adverse circumstances, and that should be a source of pride.

He worried occasionally about where ancestry-testing was headed—the risks of ethnic branding, as he called it. On the other hand, he wrote in an e-mail: Who knows, Jeff, you might be 8.35 percent Ashkenazi Jewish and related to [Secretary of State] Madeleine Albright and [Virginia senator] George Allen.

Unannounced, Beatrice Martinez Wright and her husband, Tim, arrived in Culebra. It was the fall of 2001, the aspens turning yellow on *la sierra* and the frost pinching the *Vega*. Bea brought with her a printout of the Martinez family pedigree. Males were designated by squares and females by circles, and the cancers among her relatives were marked as black bites on the symbols. The black bites almost certainly indicated the presence of 185delAG.

Bea knew the records were incomplete. Some of her relatives were missing from the chart, and she didn't have addresses for all those she was sure of. She was advised to see Maria Clara Martinez (no immediate relation) because Clara was the authority on the community, having compiled many local genealogies. Clara also was the editor and sole reporter of the weekly newspaper, *La Sierra*. Bea went to the newspaper office in San Luis.

Clara showed Bea three or four crisscrossing Martinez lineages, reaching back fifteen generations to the clan's progenitor, a Spanish American with a double surname, Martín Serrano. More important, Clara produced an in-depth chart of Bea's relatives, along with health information that she'd gleaned. She [Beatrice] was a cancer survivor, Clara recalled, and had other relatives who died, and she wanted to know how she got it, so I did her chart. But Clara was unfamiliar with the 185delAG mutation and knew nothing of any Jewish ancestors. She had heard about crypto-Jews, though. Which was no big deal since Clara doesn't get too excited about anything.

In San Luis, Bea wanted to locate her father's sister Dorothy—Dorothy Martinez Medina. Someone told her that her aunt lived in Alamosa, but to be sure, she should check with Dorothy's son Joe, who lived a few miles down the road. Look for the place where they're building a restaurant. It's just before the turn to San Francisco. Her first cousin Joseph!

When Bea drove up, Joseph and Marianne didn't know who she was. I introduced myself, Bea said, and I told them I'd just been through a bout of breast cancer. They turned white as a ghost. They said, We just lost our daughter.

The two cousins had some catching up to do. A little later, Bea said to Joseph, Well, we're Jewish.

Bea recounted the goings-on in their grandmother Andrellita's house, the covert things that she thought might have been Jewish. They were taught to stay quiet, to not say anything, Bea said. When you were growing up, she asked Joseph, did anyone ever light special candles?

Nonplussed by this turn of events, Joseph thought for a minute. He recollected that one of the aunts used to light bonfires on the nights counting down to Christmas—first ten fires, then nine, and so forth until a single fire burned on Christmas eve. Was that a Jewish practice?

(No, probably it was *Las Posadas*, an old tradition of the season. Candles, luminarias, or bonfires are lit, but in the opposite order from what Joseph remembered: one, two, three, etc., until nine lights are burning in expectation of the baby Jesus. The nights of *Las Posadas* reenact Mary and Joseph's search for an inn. Some have argued that *converso* families in New Mexico conflated *Las Posadas* with the Hanukkah celebration, since both customs featured lights, and that the nine fires stood for the nine candles on a menorah.)

Joseph retrieved another item from his memory. He had heard that in order to protect a newborn baby from a witch's spell, or was it from *ojo*, the evil eye, you were supposed to have a man named John dribble water from his mouth onto the child. Was that Jewish?

(No, basic Hispano. Men named Juan indeed were believed to have power over witches. In dribbling the water on the infant, probably they were imitating John the Baptist.)

Beatrice drove off, looking for additional family members to talk to. A year or two later, after a newspaper article on the genetic disorders of San Luis Valley mentioned Bea and her deceased cousin Shonnie, Joseph and Marianne were approached by two members of the Society for Crypto-Judaic Studies. Watching Marianne baking at the restaurant, the visitors said, Hmm, the making of bread. In the eyes of the beholders baking was a telltale Jewish tradition. Some went so far as to speculate that the tortilla was a crypto-Jewish carryover of unleavened bread, and that certain red chile dishes featured what looked to be matzo balls.

Stanley Hordes, meanwhile, decided to incorporate the 185delAG mutation into his historical research program. He was persuaded that it was a better indicator of Jewish ancestry than a Middle Eastern Y chromosome or the pemphigus vulgaris skin disorder. Broadening his territory beyond New Mexico, Hordes tracked the historical footprints of Sephardic Jews and New Christians in Peru and the Caribbean islands. I start with people today, people in those countries who acknowledge Jewish heritage or who carry 185delAG, he explained. It's a place to begin to ask questions. Then I do the historical and genealogical work.

Hordes added a panel on genetics to the annual meeting of the Society for Crypto-Judaic Studies. These meetings were lively, learned affairs. Typical topics addressed by speakers were: "Anti-Judaism and the Theology of Forced Conversions: A Textual Analysis of the Sermons of Vicente Ferrer [1350–1419]," "The Jew That Is Not One: Contemporary Theoretical Approaches and Crypto-Jewish Cultural Production," and "So Your Ancestors were Crypto-Jews: Now What? A Conceptual Model for Helping Hispanos Evaluate the Significance of their Ancestral Past and its Meaning in their Spiritual Present." To which Hordes added: "Issues in the Identification, Counseling, and Treatment of Patients with Genetic Diseases Associated with Jewish Populations," for Stan tried to keep in mind, as Harry bore in mind, that using DNA in a historical inquiry often involves people who are ill.

At the society's gatherings, the wannabes were a wild card for Stan to manage. To characterize them more kindly, these were the participants who longed to identify with Jews. Such as the Hispano woman who confessed privately, I love the Jewish people and I am a member of the International Fellowship of Christians and Jews. I would love to be Jewish, from the chosen people, even if it means persecution or a higher risk for cancer.

The emotional climax of the 2007 meeting, which was held in Albuquerque, was a talk titled simply "Crypto-Jewish Crucifix?" The speaker unfurled his story in an understated manner. During his childhood, the man recounted, there was a wooden crucifix that was kept hidden except on rare occasions. His grandmother, shushing his questions about it, would take out the piece and hang it on a bedroom wall, where it was visible for a few hours, a single day at most. Then it was whisked away.

Now that he was grown, he had come to regard the crucifix as the symbol of his family's secret faith. And as he was speaking, the man's face reddened. He started to weep. Dramatically he pulled out the crucifix, which was about eighteen inches high, and he held it before the audience. He slid the ivory-colored Christ off the wood, revealing hidden candle holes and a slot that was meant, he said, for a *mezuzah*. There were oohs and aahs all around.

Father Bill Sanchez interrupted—he the Hispano-Jewish *padre* with the cohanim Y chromosome. Father Bill said dryly, Why don't you see if the Christ fits into the slot? He was asking if the piece would stand vertically, with the cross as its base. Oops, it did. Stanley squirmed in his seat. That's just the sort of thing he hates. The artifact could be a normal Catholic devotional object, he said later, or it could be just what its owner said it was. Objects do not speak for themselves.

Early in 2009 Harry e-mailed me: So my question for you is, How would you like to be involved in research? You are the person who I know is best connected to the Hispanic community in the San Luis Valley. My efforts to collect data about them from geneticists and epidemiologists have been nonproductive. So how would you like to help me collect blood samples from the community?

We could spend 1–2 days together in NM, he sweetly continued, depending on when we went, and we could even fit in a bit of skiing. My study has IRB [institutional review board, which monitors ethics] approval and, as you have seen, I am willing to travel to inform and [obtain] consent [from] people and collect their samples. You would get a great story for which you could write a firsthand account.

To explain what Harry Ostrer was after in Culebra—he was after Jewish DNA, not to beat around the bush. But to justify this search for

Jewishness, which was different from Stanley's search, it might be good to review the scientific issues surrounding race. For Harry was interested in race and ethnicity and the restless mingling of human populations. He once spoke of himself as an Ashkenazi Jewish mutt, because even so tight a group as the Ashkenazim blended people from different villages of Central and Eastern Europe.

The concept of race is straightforward for the other creatures of the natural world: Populations of the same species that don't overlap in breeding territory can safely be called races. The geographic separation usually has caused a divergence in appearance between the groups. A race is akin to a subspecies. Race science, a term you no longer hear, was built on the assumption that the human animal had diverged into sub-species, and that the races of this species were promiscuously undermining the natural barriers set between them.

The heyday of race science occurred in the late nineteenth and early twentieth centuries, and as a serious discipline it had ended with the Holocaust. After the Holocaust, Jews and Hispanics and many other peoples of the world that biologists had counted as races were downgraded to ethnic groups. But race as a compilation of variable traits such as skin color carried on, like an ocean liner after its engines have been cut. Sociologists and anthropologists took potshots at race, peppering the phenotypes of physical difference until they were full of holes, and then the geneticists torpedoed race by showing the overwhelming similarities among human beings at the DNA level, below the waterline of the skin. Still, since cultural and socioeconomic distinctions between ethnic groups do accord with color, and since most people in thinking about race and ethnicity are rather like ocean liners, understandably the U.S. Census Bureau and other institutions have continued to recognize race. Half-sociological, half-biological, race still applies to real life.

After genomics came into play—the human genome project and the rest—the race concept made a comeback in science, although now it went by another name, continent of ancestry. Using the latest technology, geneticists found they could classify people by their continents of ancestry, which were congruent with race. If individuals were admixed, like African Americans and Hispanos, the proportions of the original, geographical DNA could be calculated. How did this new science of discrimination come about?

The human genome is so long that, even in the face of great similarity, wrinkles of difference appear. When you compare the genomic texts of any two humans, they are 99 percent identical, even greater than 99 percent. But since you have available more than three billion bases of DNA for comparison—actually six billion when you count the full complement of letters in the paired chromosomes—that 1 percent of difference looms large. Approximately twenty-four million letters differentiate one person's DNA from another's, easy enough for computers to analyze.

Before determining the geography of someone's ancestry, the computer has to have done the necessary homework on the DNA patterns of the world's peoples. These patterns started to form fifty thousand years ago, after bands of *Homo sapiens* had left Africa and migrated over the world's continents. Mating within their own spheres, humans stopped exchanging genes broadly. Entered on a graph of genetic variation, the individuals constituting a particular population cluster together. Although the same genetic variants might occur in another group half a world away, the frequencies or distributions of the variants aren't the same, and a few of the variants might be exclusive. The two populations will form separate clusters. Now peer into one of the clusters. Great similarity is again the rule, but some of the sites where a pair of human beings is identical are not by accident. Two people may have the same DNA because they have descended from a common ancestor, and on the graph of the world's populations their shared history will make them cluster together.

Ancestry-informative markers, AIMs for short, are the raw material of such studies. AIMs aren't genes; for the most part they are repetitive blocks of DNA sequence or isolated letters interspersed between the genes. A single AIM won't be revealing of a person's forebears because that one marker probably can be found elsewhere. But a panel of markers, each chosen for being common in one population and rare in another, can be used to make a reliable prediction about the race—er, continental ancestry—of the subject. In a proper ancestral study, AIMs across the entire genome are assayed, it should be emphasized, not just snippets of the Y chromosome and mitochondrial DNA, the stuff of many online testing services. Incidentally, the 185delAG mutation, although highly indicative of a particular ancestry, isn't used by scientists in AIM research. Compared with other DNA markers, this variant doesn't show up often enough in random surveys to be helpful. In most populations it's simply too rare.

As genomics programs have become more powerful and the differential markers of human DNA have multiplied, the computers do not need to refer to the geographical baselines stored in the data banks. Crunching half a million DNA sites for each individual, the programs are able to place anybody in a population cluster. Results are so accurate that they almost never fail to conform with the person's own understanding of his racial or ethnic background. If someone is an Ashkenazi Jew but denies that, the computer will be able to diagnose him all the same. Likewise for a Hispanic, although the clustering of Hispanics on the graph is fuzzy, since subgroups of Dominicans, Mexicans, Puerto Ricans, etc., overlap. The computer cannot be positive of the nationality of a Hispanic.

What if a Hispano, a kind of Hispanic from Old New Mexico, comes in with a DNA sample and claims that he's Ashkenazi? The computer will tell him he's mistaken. If the Hispano tells the computer, No, I meant to say I have *Sephardic* ancestry, making me a Jew, right?, the computer is going to have to think about it. Maybe yes, maybe no. That's why Harry Ostrer visited Culebra—to improve his profession's capacity to discriminate among racial and ethnic clusters, to bring a remote and blended population, Hispanos, into higher resolution than before, and to identify Jewish markers in Hispanos if they were there. Jewish genes having flowed into the New World under cover of the Spanish conquest, Harry wanted to know the fractions of the admixture.

He wasn't starting from scratch. In a recent study he and a team of researchers had compared the admixture proportions in Hispanics from

Puerto Rico, Colombia, Dominican Republic, Ecuador, and Mexico. All the populations combined European, Native American, and African DNA, and the allocation of the ancestral markers had been shaped by each group's colonial history, which was not a new insight to report. But the Ostrer team noticed an additional element, a tiny feature in the standard mix. Some Hispanics had Y-chromosomal markers deriving not from Europe (the colonizers), nor the Americas (the colonized), nor Africa (the smaller contribution of imported slaves), but from long-ago Middle Easterners or North Africans who were either Jews or Arabs. Here might be a sign of the crypto-Jews. Believers like Father Bill Sanchez would certainly embrace this finding, but the scientists themselves were cautious about it. Harry was hoping that the vein of Jewishness would be richer in San Luis Valley than in other parts of New Spain, thanks to the proven strike of 185delAG.

As for what the Culebrans might make of the information he was going to extract from them, Harry, unlike Stanley, wasn't concerned with identity. Their identity was up to them to work out. Both men had deep feelings about Jewish blood, but the cultural and biological aspects of Judaism were not entangled in Harry's mind the way they seemed to be in Stanley's. To Harry, genes were not identity, nor did he think that identity could be determined by genealogy. At the same time, Harry was pleased that Jewish DNA had survived in the broken populations that he had followed to far corners of the world, to Serbia, Greece, Ecuador, and now to the San Luis Valley. Traditionally, he observed, Jews have held a mourning ritual for the Jews who married out or converted, or for those whose faith was taken away by force. Harry took heart from the fact that their DNA had not disappeared, even if the markers had become separated from what they marked.

When he made his *entrada* into the Valley, driving from Denver with winter's stars twinkling over the mountains, the New Yorker looked forward to his meeting with the Hispanos. Already he felt a bond with these folks, stemming from their shared social experience, not their DNA. For in striving to overcome prejudice and become good Americans, the Hispanos were not unlike the Jews. They feel they have a place at the table, Harry explained. They've claimed a place at the table. I recognize them at a certain level. The older people—he was speaking of the European immigrants from his own tribe, but it applied to the Hispanos also— tend to feel they are just guests in someone else's country. The younger people don't feel that.

He made his way through the dark village of San Luis to his room at El Convento. Tomorrow's sampling session had been billed on posters and in the local paper as the Hispano DNA Project, a basic admixture study, sidestepping the role of crypto-Jews. Yet Harry was confident that Jewish ancestry wouldn't faze the Culebrans. These people are already hybrids, he said. They are comfortable with being mongrels. He meant nothing pejorative by that, because Harry was trained to pass through the firewalls of race and ethnicity and access the DNA that each person had downloaded, so to speak, from scattered, ancestral servers. As a medical geneticist, he was well aware that where your DNA comes from can have implications for your health, and that racial or ethnic labels can help guide physicians. But race was only a way station, just as Harry's admixture studies were only a means. Someday fairly soon, when the technology and the cost allow it and the hype of the DNA age has died down, the genomes of every individual will be known for their unique pluses and minuses, and then doctors won't need categories like race and ethnicity.

February 21, 2009, was a brisk, clear, dry Saturday in Culebra, with scant snow on the ground. Marianne Medina got up at sunrise. After helping to distribute the posters earlier, this morning she drove up and down the country roads, tying bunches of blue balloons to telephone poles and signposts. Balloons were a custom in the community to reassure drivers that they were going in the right direction—to T-ana's Restaurant, the host of the Hispano DNA Project, which had come together because of her daughter, dead ten years ago this winter.

Shonnie's cousin Beatrice Martinez Wright had a bad case of the flu, but Bea insisted on making the long drive from northern Colorado. Bea, her husband, and their son stayed on Friday night at the San Luis Inn Motel, Culebra's only motel, which intercepted visitors who might otherwise veer into the bed-and-breakfast at El Convento. Harry arrived at the motel around 9 a.m. with his medical bag. Fittingly, Bea would donate the first blood sample. She wasn't up to attending the group session.

While her husband and son hovered in the background, the doctor explained the procedures to Bea, who sat primly on the side of the bed, pale but game. When she had signed the consent forms, she held out her arm, and Harry drew three vials of blood, about 20 cc in total. It seemed an awful lot to take from a breast-cancer survivor who wasn't feeling well.

Besides Bea, several members of the extended Medina-Martinez family had been invited to the community meeting. The advantage was to salt the group with genetic carriers who, because of their BRCA markers, were known to be descended in some way from Jews. The disadvantage was that the sample was not scientifically random and would not necessarily produce a representative picture of Hispano DNA. Whoever showed up had been pulled in by advertising and word of mouth, with Bea herself helping to spread the word about the meeting to her relatives. Who knows exactly what we will find? Harry said, acknowledging the uneven way that subjects were collected. He parked his car behind T-ana's. Marianne's small lot was filling up already.

Stanley Hordes, traveling with his wife, got lost and missed the 10 a.m. starting time. Cars lined the side of the road, and he was surprised by the turnout. This middle-of-nowhere café, Stan remarked. I was not prepared, he said, for the number of people who were willing to open their veins. About forty people were inside the restaurant having coffee and pastries, and thirty more were on the way, coming not only from the Valley but also from Albuquerque, Pueblo, Denver, all returning to the place of their roots. The *Pueblo Chieftain* sent a reporter, and a genetic counselor, Kate Crow, arrived from the Springs. It was a good thing Harry had decided to hire a local phlebotomist to help him. Marianne and Iona were busy in the kitchen.

Questionnaires and consent forms were stacked on a side table. The one-page questionnaire asked participants to list the names of the past five generations of their relatives, along with their relatives' places of birth and death. A complete family inventory reaching back that far would need to contain dozens of names. But here only the paternal and maternal lines were to be entered, the father's father and mother's mother, etc., for five generations. This type of family tree, which limits itself to the two enclosing branches, is designed to support Y-chromosome and mitochondrial DNA analyses.

The consent form, ten pages of boilerplate, required the study subjects to read and sign their initials on each of the pages. Harry was very serious that people should know what they were getting into when they enrolled in genetic research. Taking a microphone to the far end of the dining area, where Joseph used to perform, Harry walked the participants through the issues. The main thing they must understand, he said, was that no individual results would be reported. Their names would be stripped from the samples—anonymized, as he put it—and only a collective portrait of the DNA would be given back to them. This was meant to smooth out idiosyncrasies and prevent anyone from worrying about individual health risks. You will get information back about your community, Harry stated. Though we are interested in the genetic basis of disease, we are not approaching this study as your health-care provider.

Highlighting the ambiguity of the Hispano DNA Project, the consent form was the same one that Harry used in his Jewish ancestry project at NYU. The top sheet was headed "Origins and Migrations of Jewish Populations," and a summary section read: "The purpose of this study is to trace the origins of the Jewish Diaspora and to examine the genetic similarities of Jewish peoples, including Cohanim, Levites, and Israelites from the various ancient Jewish tribes." The text advised: "You have been selected because you are a Jewish person with knowledge of your geographical ancestry." In language that was a bit too technical for the audience, Harry explained how the larger study was going to be tweaked for the Hispano population. The purpose of this study is not to demonstrate that you all are Jewish, he said, while reminding them about the 185delAG mutation and its link to Spanish *conversos*. But as Harry had figured, the ambiguity of his message wasn't a big deal. Eager to learn about their genetic heritage, the Hispanos consented without quibbling, and the only person to get up and leave was a man who had expected to receive an individual test report.

Dr. Ostrer held up the three tubes that needed to be filled; two had purple tops and one had a yellow top. Why take blood and not saliva? asked Clara Martinez, the newspaper editor and genealogist. DNA is of higher quality in blood, Harry replied, adding that the samples would be maintained in special cell lines in the laboratory. This step would immortalize their DNA, he might have pointed out, so that their genes could be studied indefinitely.

Clutching their paperwork, the Hispanos advanced in stages from the restaurant tables to what might be called the waiting room, a bank of chairs up front, and from there to the phlebotomy station, a small table manned by Harry and his hired helper. Behind them was a local nurse, a last-minute volunteer, who took the samples and packaged them with the completed forms. And in the back corner, behind the medical team, Clara set up her computer containing the names of thousands of New Mexicans of the past. Although several families brought along their pedigrees, many others were able to fill in the gaps in their histories by consulting Clara's genealogies. Throughout the session, there were hiccups of laughter and surprise at what she was able to dig up: a long-lost grandparent, or an uncle who was better off staying lost.

Here comes Maclovio Martinez, his sleeve rolled up. Maclovio is a fierce and modest traditionalist of Culebra. Here is Ruben Archuleta, the self-published author of a book on New Mexico's *penitentes*, copies of which Ruben has brought with him. Here's the ebullient Debbie Rich-Crane, who was last seen at the family counseling session eighteen months ago. She was Jeff Shaw's demonstration wife, who used soda cans to illustrate the inheritance of the 185delAG mutation. Another of Shonnie's first cousins, once removed, Debbie is very keen on her Jewish-Christian heritage. I left the Catholic Church as a teenager, she wrote in an e-mail, and am now a born-again Christian and know that the Jewish people are God's CHOSEN PEOPLE! Wow, how great is this!!! She signed her e-mail A2J, short for Addicted to Jesus.

Next up is a teenager in a puffy pink sweatshirt named Najondine, who's from a different branch of Shonnie's family. She is with her anxious, pretty grandmother, Kathy. But Marianne Medina and Iona do not give a blood sample, maybe because they are occupied with serving food to the people at the back of the line, or maybe because the two just don't want to participate. Iona will go to Colorado Springs for her BRCA test a few months hence, following her Auntie Wanda's death. Yes, Wanda Kramer and her husband, Bill, are at the phlebotomy station now. Being Anglo, Bill can't contribute blood to the study, but Wanda steps up willingly with her big, warm smile. When she turns and asks Bill her birth date, her forgetfulness seems strange. Her body and possibly her brain are riddled with cancer.

Harry was feeling great, he admitted afterward. Having brief, animated conversations with everyone who sat down, asking why they have come and what they hope to learn, not slowing as he maneuvered the syringes and capped the somberly glowing vials of blood, and then handing off the samples rapidly, without a glance behind him, just as on that pioneering day in 1975 when he tested hundreds of Jews in New York for the secret signature of Tay-Sachs disease.

Stanley Hordes, not to be outdone, worked the room. Seemingly without prompting, people would come up and tell him fragments of their pasts that could be construed as Jewish. True, he and his book had been introduced to the group at the beginning of the meeting. Some who were there knew about their Jewish heritage, Stan recalled. Some were comfortable with it. They'd say, There were always stories that our family used to be Jewish. Other people told me, You need to go talk to so-and-so.

During the lunch break Stan and Harry sat at a table together, their first occasion to meet each other. They traded information, casually showing their bona fides. Harry knew more about Jewish culture than Stan knew about Jewish genetics, but Stan would not be ruffled and pressed forward in his dogged manner, while Harry, who can be skittish with people he doesn't know, turned aside some of Stan's queries with a polite flutter of the eyelids. It was fun to see them together and hear them exchange stories. Both were good at wisecracks, Harry's cutting deeper.

The blood-draw session wound down around 4 p.m. Harry and his assistants worked another hour going over the forms and assigning identification numbers to the samples, and as they worked, Iona and Bill sang karaoke songs toasting the success of the day. Stanley and his wife drove home to Albuquerque. Harry had a margarita and dinner at Lu's Main Street Café in Blanca, up the road from Culebra, at the foot of the magic mountain, and afterward he went to the Walmart in Alamosa to buy a Styrofoam box for his DNA trove. At dawn the next day, he left the San Luis Valley, having done no skiing and no sightseeing except for a brief tour of the Stations of the Cross Shrine and the San Francisco church. He never said if he liked staying at El Convento.

The processing of the samples took time. In the summer of 2010, Harry sent a letter to the study participants with his preliminary analysis.

"The question of greatest interest to the community," Harry wrote, "and the first one we tried to answer, is about the proportions of admixture. From testing your DNA we found that, on average, Hispanos are:

- European 50–60%
- Native American 30–40%
- West African 1–5%
- Non-European (Middle Eastern) 1–5%."

The first three numbers were in line with previous studies of Hispanic groups. The detection of a Middle Eastern component in Hispanos was new.

But behind the scenes, Harry was a little disappointed because Jewishness had not jumped out of the DNA samples. Was it a question of not looking hard enough, he wondered, or not applying the right tests or markers? Or not getting the right ideas from his collaborators? Under the umbrella of his master project, "Origins and Migrations of Jewish Populations", Harry tended to work inductively, feeling his way through the dark corridors of the data toward a scientific conclusion. He wasn't really joking when he said that he liked to do the experiments first and then find a hypothesis that fit the results.

It's like a journalist, he explained, who has to find the story. What's the best way to tell that? What's the story line? The Ostrer lab also was like the TV show *House*, with Harry in the role of Dr. House peppering his younger associates, making them sweat as they reached for the right diagnosis. We need new insights to improve the story, Harry said.

He decided to compare the Hispanos with a second group of Hispanics. In the Loja province of the southern highlands of Ecuador there lived a curious group of villagers, the Lojanos, who carry a recessive disorder called Laron syndrome. The effect of the gene is to severely stunt growth and cause other health problems. The mutation has a Jewish signature, presumably because *converso* ancestors of the carriers fled Lima in the seventeenth century, when the Inquisition was active. What's more, Harry had been told that in the affected communities some people avoided pork and lit candles on Friday night, the kinds of stories Stanley heard about crypto-Jews in New Mexico. Although no one in Loja has claimed to be Jewish today, Laron syndrome in the inbred Andean colony might be a thread that if tugged would uncover a new trail of secret Jews in America.

So Harry obtained DNA samples from the local investigator of the Laron condition. He submitted the Hispano and Lojano DNA to more rigorous analyses than before and found his story line: an intriguing genetic overlap between two groups of Hispanics located thousands of miles apart. At least ten men had Middle Eastern Y chromosomes. When the Hispanos and Lojanos were compared with European groups, including Sephardic Jews, the two New World populations clustered closer to the Sephardic Jews than did any of the comparison populations.

The shared DNA, thanks possibly to a Jewish nexus in their past— that was the story that Harry and his colleagues published in the journal *Human Genetics* in 2011. The title of the article was "The Impact of *Converso* Jews on the Genomes of Modern Latin Americans."

Harry's scientific articles made news. He finished work on a book about Jewish genetics and helped organize a conference on the subject in Israel. With his career on the upswing, he moved from NYU to the Albert Einstein College of Medicine in the Bronx, New York, joining forces with high-powered specialists and becoming the director of the DNA laboratory there. Given that the Bronx's population was half Hispanic and 6 percent Jewish, he planned to do more work on the Jewish origins of Latin American populations and to develop DNA tests of medical benefit to both groups. He turned sixty without slowing down or looking back.

After you have been a study-partner with someone for seven years, you don't expect surprises at the end, but Harry did surprise me with a personal story about 185delAG. He too was touched by the gene. The mother of his longtime friend A—— had died of breast cancer when she was in her early forties. My life was changed by this woman's death, Harry said, because afterward A—— was always around during his vacations from boarding school. He became very close to my parents and to my extended family. My life was enhanced by A——'s friendship.

Much later, a sibling in A——'s family, concerned about cancer, learned that he was positive for 185delAG. The test result not only explained the mother's swift disease but also raised questions for A—— and his daughter, who found themselves in the same shaky boat as the Medinas and the Martinezes, two more people squeezing the rail of DNA analysis and looking over the side. They wanted to know yet dreaded to know.

The lesson, Harry said, is that genetics is an equalizer across socioeconomic lines. These [BRCA] mutations are common enough that they can have an impact. These people's lives were changed, he repeated—meaning Shonnie's Hispano family, A——'s Ashkenazi family, and Harry himself. And it wasn't just a fortuitous or random inheritance of genes, Harry said. There are reasons for what happened. There are historical reasons.

Chapter 10

THE OBLIGATE CARRIER

*And Aaron shall lay both his hands upon the head of a live goat
and confess over it all the iniquities of the children of Israel—
all their transgressions, all their sins, all put on the head of
the goat. Then he shall send it away into the wilderness by the
hand of a ready man.*

—LEVITICUS 16:21

Before she discovered makeup, hair curlers, and the startling impact of her beauty, Shonnie Medina was a tomboy. She and Rod Gallegos, a childhood playmate from the Kingdom Hall, would chase sheep in the *Vega*, the common pastureland of Culebra, running after the annoyed beasts all afternoon. When Shonnie got older and learned to ride, she'd gallop on Hot Smoke through the same fields, or take off for hours into the foothills of *la sierra*.

She and Iona were happy go-getters, her aunt Lupita remembered. They were always wanting to learn things, and Joseph was always teaching them how to do things. He didn't have boys but it was like they were boys. Always riding and fishing, things of that nature.

She grew up bucking bales, her aunt Wanda remembered. Shonnie looked totally different when she was on the farm, wearing jeans and a baseball cap. She'd stuff her hair into the ballcap or put it in a ponytail. Later, when she was married, Shonnie was an equestrienne in full makeup who enjoyed teaching her young nieces to ride. Family friend George Casias said, I was privileged to see her at her peak—when she was being outdoorsy. She loved dogs and her horse.

Outdoorsy Shonnie was bolder and handier than the feminine Shonnie let on. Some of those moments when she was ditsy or klutzy might have been feigned. When she was nineteen, she and her sister went on vacation to Hawaii along with some other teens from the Jehovah's Witnesses. Though she couldn't swim, Shonnie got up on a rock with the boys, fifteen feet above the water, and jumped off, making a tremendous splash. After a frantic dog paddle to the shore, she climbed up and jumped off again. She was very determined, said Iona. She would do things by herself if no one wanted to come along.

If the showy, headstrong person of the previous chapters was modeled on Marianne, the outdoorsy, self-reliant Shonnie was the creation of her father. Joseph Medina was proficient at carpentry, plumbing, adobe-making, logging, lumber-milling, woodworking, masonry, and engine repair, not to mention guitar-playing and singing. Joseph hoped that his girls would acquire as many of his skills as possible, in case something should happen to him and they had to go into the world alone. But of course they inherited more than that; their lot was as unlucky as his in regard to the 185delAG mutation. The gene that had passed from Andrellita to Dorothy to Joseph was transmitted to Shonnie, the one thing he couldn't protect her from. He never found out that Iona was positive too.

After Shonnie died, Joseph took to calling Iona his only heir. Most of the property that his father had owned belonged to Joe Jr., for he had bought out the shares of his younger brothers. The eldest son felt a responsibility to the legacy of Joe U. Medina. But his own hand-built home was already given over to Iona and her husband, and she stood to get all the land and the restaurant building too. As for the money he'd saved during his life, Joseph did not trust the banks to hold it. He buried jars of cash on his property, careful not to use metal lids since the lids might be picked up by a thief's metal detector. To mark the locations of the stash, he gave

Marianne sketches of trees and other objects. One of his drawings pictured a beaver-chewed log, which took her weeks to recognize.

Costilla County, where the Hispanos are clustered, is the poorest county in Colorado, according to government statistics. For the three thousand–plus residents, the median household income in 2008 was $25,000, and 10 percent of workers were reported to be unemployed. But if Joseph's finances were a barometer of the economy, barter, an activity without cash, was a significant part of the picture, and other earnings were being squirreled away under the mantle of self-employment. For as long as Anglos have been in New Mexico, they have complained that Hispanos did not care about making a dollar. T-ana's Restaurant, Marianne's domain, had a modest cash flow, but the freewheeling Joseph Medina, who waited on tables and sang on the weekends, had no ambition to amass or consume money. He was not lazy. He worked to survive, to be his own boss, to be in tune with the seasons of life and to connect viscerally to the land as his ancestors had.

Joseph's landscape commenced behind the restaurant, on the far side of a fence made from lashed-together saplings. There was a shed with a small sawmill, some well-worn farm equipment, and then hayfields, which he was preparing to harvest. His land extended east to a spring-fed pond containing a few trout. On summer evenings after work Joseph would cast for fish, clambering through the willows on the banks of the pond and trying not to fall in. Like Shonnie, he couldn't swim— too bad, since the water was great, with a fine floating view of Mount Blanca. Proceeding east, you left Joseph's property, yet the landscape was still his own, because historically the Hispanos' rights had extended to the top of the Sangre de Cristos, a mile into the sky from the valley floor and covering probably fifteen or twenty miles of terrain between his house and the crest. Over the past century and a half the access to the resources of this land—its timber, game, water—had been denied and then partially restored to the descendants of the original *pobladores*. Today Joseph Medina was going to make an expedition to the summit of *la sierra*, a little lord reclaiming his right to cut firewood. But before leaving, he had to bale and stack his hay crop.

Late August 2007, a shimmering day in Culebra. Cut and raked, the hay lies in thin rows on the green stubble. In the neighboring lot, two big-bellied men on a wagon are handling bales thrown up by two younger guys who follow on foot. A couple of magpies, black and white as if cassocked, are fluttering, hopping, foraging at the edge of the field. The lazy crow of a rooster somewhere. Hollyhocks blooming, both red and white, the wine and the wafer. The redolence of hay on a summer afternoon, and the more delicate, intermingling wafts from the great untended prairie to the west. Novelist Willa Cather had responded deeply to the hybrid smells. "[T]he lightness, that dry aromatic odour," she wrote of the prairie. "The moisture of plowed land, the heaviness of labour and growth and grain-bearing, utterly destroyed it; one could breathe that only on the bright edges of the world, on the great grass plains or the sage-brush desert."

Joseph usually brought his hay in by himself. Marianne could help, but was less cooperative since the time her foot got caught in a trailing strand of baling twine. She was dragged many yards, her screams unheard over the noise of the tractor and the chugging baler. Since he had hurt an arm and shoulder a few years ago, Joseph worked with a smaller wagon and slung up the scattered bales with his good arm. Obviously it would go faster if he drove the tractor and someone else picked up the bales. A ready man was at hand, but Joseph hated asking for assistance. If you want to, he'd shrug, looking away. Soon enough the work in the field was done. A rancher and his sons arrived and took all the hay, paying Joseph cash.

After lunch he took his battered truck on the county road east toward San Francisco. The rabbitbrush was yellow in the haze. Through the village, past the forlorn church and the half-hidden *morada*, the road climbed into the foothills of what used to be called the Culebra Range. My dad took us there as kids, Joseph said, glancing at the *morada*. In his speech he had a slight, stop-and-go syncopation, from the Spanish. There was a private room, I remember. . . . It was too scary for me as a kid.

Why did you take your own kids then?

I took the kids because of my dad, he said. One year during the Passion procession, he had carried the cross to the little *calvario* behind the *morada*. This information came from Marianne, since Joseph volunteered very little. Two or three sentences in a row was a lot for him.

Joseph Medina had mellowed. In a photograph from his younger days, which shows him roofing a house, he's wild-haired and bare-chested, his arms pop menacingly, and he looks like he'd be happy to climb down and rip you a new one without any provocation. He was a brawler, he admitted, and was known for having a bad temper. But when I fought, Joseph said, I would also be the protector of kids who got picked on. If no longer trim, Joseph was still plenty muscular, and being short had never been a problem for him. Oh, you want to go a few rounds with me? he'd offer, not quite kidding, the old reckless gleam flashing in his eye. Around guys he'd grown up with, sometimes he would growl and pretend to be angry, just to see them flinch. The rough fun of Culebra.

By the age of fifty-six, a man has learned a few things about life, or else life has forcibly taught him. Age, marriage, a religious conversion, the death of a child—the chippy self was gone, and a sadder and more humble Joseph went up into the mountains. He never finished grieving for Shonnie, his sister Chavela said. It hit him like a ton of bricks, agreed family friend Celina Gallegos. He didn't recover from it. I think it was a grief he couldn't stand. And now that he knows that I have it, put in his sister Wanda, he's scared for me, poor guy.

Shifting on its springs, the truck passed Rael Road, which runs down and dead-ends at San Francisco Creek. The surname Rael is not unusual in Hispano territory. Investigators of crypto-Judaism notice when they hear it, though, for Rael is thought to have been shortened from Israel. Joseph didn't know if any Raels were living there now.

He came to the bars of a swinging metal gate, got out, unlocked it, and locked it behind him after he had driven through. From here to the top of *la sierra* was private property owned by an Anglo named Hill, who was the latest in a long line of Anglo landholders. The original proprietor of the Sangre de Cristo Grant, Charles Beaubien, provided deeds to the Culebra settlers in 1863. Describing both private lots (*extensiones*) and public lots (such as the *Vega*), the deeds included permission to gather wood and graze livestock in the uninhabited uplands. These rights were curtailed when land companies bought up the mountain tract. In the early 1900s, both parties to the dispute filed lawsuits and title actions. The Hispano custom of sharing resources perplexed the American judicial system, which repeatedly sided with the wealthy advocates of private property.

In 1960, the mountain tract, now whittled down to some eighty thousand acres, was taken over by a redheaded lumberman named Taylor, who fenced off and patrolled the gravel roads, leading to incidents of gunfire and racially tinged vigilantism. The Taylor family sold and another owner came and went; the legal case turned into an American *Bleak House*. At last, in 2003, a lawyer from Denver named Jeffrey Goldstein, who'd worked without charge for twenty-seven years, convinced an appeals court to restore the access rights to the heirs of the settlers, provided they still lived in Costilla County and could prove their descent from an original title-holder. Gate keys were issued to three hundred tenacious Culebrans, among them Joseph Medina.

Up on the mountain, the piñon, juniper, and scratchy vegetation of the foothill zone gave way to the bigger trees of the mixed-conifer zone: spacious ponderosa pines appeared, with their handsome checkerboard bark; white-barked fir, blue spruce, and graceful, pale-skinned aspens; and now and then a Douglas fir. The vehicle bounced across small, braided streams at nearly every bend in the steep, stony track. Tucked into the dark forest, the mountain ash flashed its bright red, inedible berries. Not even the bears eat 'em, Joseph remarked.

Occasionally he would stop to point out fruits on the bushes and plants that people traditionally gathered for food and medicine. Wild rose hips (*champe*), for example, a fruit with an enormous seed, and a spray of ripe raspberries. He got out to pick some. A chipmunk with a mushroom in its mouth scampered by, pleased as punch. This is our Walmart, Joseph said. We lived off the mountain and we fended for ourselves. After Taylor fenced it off, we had to come here and poach our food. . . . Now that we've got our mountain back, we can survive.

If you killed a deer, he went on, you would hang it in a cool cave so the meat wouldn't spoil—he knew where the caves were—and you'd come get it later. When he was younger, Joseph got into canyons up here where no one had ever been. Some of his backwoods lore had a secondhand feel, such as his story about old-time sheepherders who would hang their game forty-five feet up in a tree—forty-five feet, he stipulated, just above the range of the flies. Joseph had started to expatiate, you see, and was becoming less like himself and more talkative as he went higher. Squatting, dipping a tin cup into a pine-scented stream, looking about for wildlife, grabbing deep breaths from the thinning air, the vessel of his family's pain imbibed the wilderness.

Hill, the current owner of the mountain tract, employed Joseph during elk season as a hunting guide. Men from Texas and New York flew in for a few days every fall and Joseph would take them to places where they could get a sure shot. Elk hunting took place near the tree line, at the edge of the spruce–fir zone around ten thousand feet, where the animals would come out in the open late in the day. Wearing his Western hat, mustache, and boots, his cell phone on his belt, Joseph would play the part of the affable hunting guide. The guide reaches in for the steaming heart and liver while the sportsman stands by admiring the trophy rack. Once or twice Joseph referred to rich people and to Hill's ranch as a rich man's playground, but no envy was apparent. These were simply facts.

The Culebra mountains had a raggedy look on top. Since the previous owner had logged the stands of pine, in the 1990s, there were skinned patches on the shoulders of the peaks. With characteristic ambivalence Joseph said, They raped the mountain. . . . But logging and thinning restores the big trees. A temporary benefit of the logging was that the aspens, so brilliantly yellow in fall, had invaded new areas. Also, the heirs who came from Culebra to collect wood or wild plants could make use of the logging roads.

After parking the truck next to a pile of slash, he gave a couple of toots on his elk call. Nothing was stirring in the August heat. Elk-hunting time was October. He took out his binoculars and scanned the slopes and gorges. In October they'll be everywhere, they'll be coming out of the woodwork, he said. On foot, Joseph climbed higher, steeply into the alpine tundra and boulders, his thighs brushing by the last stunted specimens of Engelmann spruce. The afternoon clouds threw oblong shadows across his path, and the whole of San Luis Valley spread out beneath him. Turning around, you could see a white blaze on the mesa in San Luis far below—the Stations of the Cross Shrine.

After a few minutes Joseph gasped, having to sit down. He couldn't go any farther. Your strides are longer, he said. The color of the boreal carpet was chartreuse. It swept upward to a band of gray shale at around twelve thousand feet—and still the rock went higher, until scraping the patches of blue.

A pika, the cute little rodent of the crags, squeaked piercingly. A pika shows itself in profile so that its omniscient sideways eye can keep you in sight. In this severe place you feel stripped down and open. When monotheistic religion began, it had only three elements, recapitulated here: Man, the unforgiving material of Earth, and God. That was enough for Judaism to work with, but Christianity added an intercessor, a scapegoat to bear the weight, who was Jesus. If, light-headed and lofty with such thoughts, you focus on the pacing of your boots, you realize that the footing is surprisingly spongy because the moisture from the winter has never departed, and the tiny plants, entwined in a mat, cushion the way between you and the rock.

A pair of northern goshawks, dark against the sky, floated over the top of their range, then abruptly tumbled through the air, swooping on each other. A pika's ear is whorled like a marine creature, like a nautilus. Marine fossils can be found on top of *la sierra*, Joseph noted. He was waiting contentedly on a lichen-spotted boulder near the treeline. As a Jehovah's Witness, he was inclined to take mountain fossils as evidence of Noah's Flood rather than of tectonic and evolutionary processes. Joseph lived in a scripted, ordered universe, not too different from the medieval Catholic universe, where the sublunary struggles of human existence rose zone by zone into the realms of perfection and peace. That said, Joseph really was very tolerant of other people's understandings, as Jehovah's Witnesses tend to be.

It was time to head down, since thunderheads were massing. The *kat-*

sina spirits are welcome except for the lightning they cause, which can be dangerous on the heights. Joseph had brought a chain saw for the firewood he meant to gather. When the truck got stuck in a wet spot halfway down, he cut the wood there and put it in the bed for weight on the rear wheels. In due course he arrived at the gate, unlocked it, and locked it behind him.

A gorgeous field of tall sunflowers was growing near Rael Road. Although they looked like they must be someone's crop, they were wild sunflowers on uncultivated bottomland by the creek. Joseph said that his father had owned that property for a brief while. When the man his father had bought it from changed his mind, Joe U. generously agreed to sell it back.

He pulled over and walked into the field, hundreds or perhaps thousands of yellow heads nodding above his. As the clouds writhed and darkened, he thought of an article he had read about environmental experiments conducted on sunflowers. Harnessing the plants' vigorous growth, scientists have planted sunflowers in contaminated soil, where they draw the unwanted chemicals and metals into their tissues. Poor people, who haven't gone to school—they knew that sunflowers can detoxify soil, Joseph said. Just as his radiant Shonnie had done by sucking the bad from people's hearts and shining the good from her own.

As soon as his truck was moving again, Joseph turned off the engine. He let the vehicle coast through San Francisco. He always did it on this road to save gas, he said. Silently he rolled past the *morada* and the church. The weary gray face of Mount Blanca did not move from the horizon to the right. Passing fields and *acequias*, the truck hardly had to brake because the road had no traffic. After three or four miles of downhill running he came to an intersection and a stop sign, a short distance from T-ana's Restaurant.

Joseph looked over with a small smile. Might as well spend some of that Jewish money, he said, and turned the ignition key.

The plan from the moment he made that quip was to end this book there. But just two months later, guiding an elk hunt, Joseph Medina sat down in the woods and toppled over, killed by a heart attack. That morning, Marianne had asked him to bring her aspen leaves from *la sierra*. She wanted some cheerful aspen leaves for her vase. His funeral service was held at the Kingdom Hall on October 20, 2007.

Marianne had three sets of ashes. She had kept Shonnie's ashes all along and now she had Joseph's ashes, and also her father-in-law's, which had been in Joseph's custody. She mulled over what to do with them. Wanda's death, in the spring of 2009, spurred her to act, because the Medina brothers and sisters would be gathering in Culebra for Wanda's funeral.

In late June, following Wanda's service, the family members went up onto *la sierra* in their trucks and four-wheel drives. They got permission from Mr. Hill to visit a picnic spot, once the site of an old corral with good grass and a creek. Medinas and Martinezes had enjoyed it for generations, when they had access. We'd camp there when Shonnie and Iona were kids, Marianne said. They rode their horses up there.

Marianne was in charge and didn't discuss with her relatives that there would be no ceremony or remembrances during the spreading of the ashes. We [Jehovah's Witnesses] don't do no prayer, no nothin', she said.

Opening the three boxes and taking out the three plastic bags, she marveled at the distinctly different qualities of the remains. The grandparent, Joe U., had made ashes that were dark and crumbly. Her husband's were gray, solid, and very heavy. Shonnie's ashes were fine and nearly white.

They did Shonnie's first. Shannon took the first turn with the ashes because she had been so close to Shonnie. Next Joe U.'s, spread by one of the sons, who cried. Then Joseph's, with Dorothy breaking down as she sprinkled the burnt genes of her eldest boy on the landscape. Taking her turn last, Marianne finished up. That was Marianne Medina for you. She came along after people who were flagging and who needed help, and she finished up.

ACKNOWLEDGMENTS

My profound thanks go to the J. S. Guggenheim Memorial Foundation for awarding me a fellowship and providing vital support for this book. Thanks also to my intrepid literary agent, Lisa Queen; and to my editor, Starling Lawrence, for seeing the value of my work, past and present.

This project was ten years in the making, if I count the time since I started to write about genetics. A number of people helped along the way, doing more than was asked and/or inspiring me. More or less in order of their service, they were: Ted Friedmann, Steve Petranek, Sarah Richardson, Victor McKusick, Brad Margus, Wayne Grody, Ed McCabe, Ginger Weber, Georgia Dunston, Leena Peltonen, Neil Risch, Funmi Olopade, Lawrence Brody, Carey Winfrey, Terry Monmaney, Teresa (Tess) Castellano, Lisa Mullineaux, Jeffrey Shaw, George Casias, Ricardo Velásquez, Maria Clara Martinez, Paul Duncan, Mike Multari.

I am especially obliged to two dedicated men who shared their research and patiently answered questions: Harry Ostrer and Stanley Hordes. Many others I don't have space to name responded to my queries, for which I am grateful. I thank my steadfast, beautiful wife, there at the beginning and at the end.

Finally I must acknowledge the members of the Medina and Martinez families. They were always gracious, forthcoming, and kind, even as their losses continued. I admire you and wish you peace.

NOTES

PROLOGUE

PHOTO: Mount Blanca, San Luis Valley, Colorado. All photos by the author unless otherwise noted.

2   *Hispano*, the Spanish word for Hispanic, is a term used by historians and social
scientists. It denotes people whose forebears lived in what became U.S. territory
after the Mexican-American War. Most other professionals, including medical professionals and geneticists, tend not to distinguish Hispanos from the larger class of
Hispanics. Hispanos themselves usually say they are Spanish or Spanish American,
and sometimes they will call themselves Mexican, but only when they are speaking
in Spanish, not English. In English, Mexican is reserved for later immigrants.

     As for Hispanos' perceptions of their Native American blood, see "Genetic
Admixture, Self-Reported Ethnicity, Self-Estimated Admixture, and Skin Pigmen-
tation among Hispanics and Native Americans," Yann C. Klimentidis et al., *American Journal of Physical Anthropology*, Vol. 138, 2009, pp. 375–83.

6   My edition of *Don Quixote* is the 1755 translation by Tobias Smollett: *The Adventures of Don Quixote de la Mancha*, Miguel de Cervantes, Farrar, Straus and Giroux,
New York, 1986. The discussion of Dulcinea's origins, which I have partly quoted
but mainly paraphrased, appears in Volume I, Book 3, pp. 190–92.

CHAPTER 1: GIRASOL

PHOTO: Wild sunflowers, San Francisco, Colorado.
EPIGRAPH (Spanish verse only): *The People of El Valle: A History of the Spanish Settlers in
the San Luis Valley*, Olibama Lopez Tushar, El Escritorio Press, Pueblo, CO, 2007.

CHAPTER 2: PREDESTINATION

PHOTO: Stations of the Cross Shrine, San Luis, Colorado.
27   For information on sporadic breast cancer, see the American Cancer Society guide:
http://www.cancer.org/Cancer/BreastCancer/DetailedGuide/index.

     For BRCA inheritance and penetrance I consulted: "Breast Cancer Risk Associated with BRCA1 and BRCA2 in Diverse populations," James D. Fackenthal and Olufunmilayo I. Olopade, *Nature Reviews/Cancer*, Vol. 7, December 2007, pp. 937–48; "Cancer risks among BRCA1 and BRCA2 mutation carriers," E. Levy-Lahad and E. Friedman, *British Journal of Cancer*, Vol. 96, 2007, pp. 11–16; "Hereditary Breast Cancer in Jews," Wendy S. Rubinstein, *Familial Cancer*, Vol. 3, 2004, pp. 249–57. See also "Genetics of Breast and Ovarian Cancer," by the National Cancer Institute: http://www.cancer.gov/cancertopics/pdq/genetics/breast-and-ovarian/HealthProfessional.

A good book for the lay reader concerned about heritable breast and ovarian cancer: *Positive Results: Making the Best Decisions When You're at High Risk for Breast and Ovarian Cancer*, Joi L. Morris and Ora K. Gordon, MD, Prometheus Books, New York, 2010.

32   A catalog of BRCA1 and BRCA2 mutations is maintained by the National Institutes of Health, accessible at: http://research.nhgri.nih.gov/projects/bic.

The 185delAG mutation of BRCA1 also goes by the name 187delAG. The former entered the literature first, but for technical reasons many scientists have adopted the latter. "The naming confusion is a result of the fact that the sequence from nucleotides 185–188 is 'AGAG,' so that it is impossible to tell if the deletion is caused by removal of the nucleotides at positions 185–86 or at positions 187–88, as both would produce the same final sequence." Source: http://www.pharmgkb.org/search/annotatedGene/brca1/variant.jsp.

CHAPTER 3: THE WANDERING GENE

PHOTO: Culebra Range, Sangre de Cristos.

37   I found the following surveys of Jewish history and Jewish genetics helpful: *Wanderings: History of the Jews*, Chaim Potok, Knopf, New York, 1978; *The Chosen People in America*, Arnold M. Eisen, Indiana University Press, Bloomington and Indianapolis, 1983; "A Genetic Profile of Contemporary Jewish Populations," Harry Ostrer, *Nature Reviews/Genetics*, Vol. 2, November 2001, pp. 891–98; "The Jewish People: Their Ethnic History, Genetic Disorders and Specific Cancer Susceptibility," Inbal Kedar-Barnes and Paul Rozen, *Familial Cancer*, Vol. 3, 2004, pp. 193–99; "A Mosaic of People: The Jewish Story and a Reassessment of the DNA Evidence," Ellen Levy-Coffman, *Journal of Genetic Genealogy*, Vol. 1, 2005, pp. 12–33; *Abraham's Children: Race, Identity, and the DNA of the Chosen People*, Jon Entine, Grand Central Publishing, New York, 2007.

40   For an example of a DNA test for Jewish ancestry, see: "A Genome-Wide Genetic Signature of Jewish Ancestry Perfectly Separates Individuals with and without Full Jewish Ancestry in a Large Random Sample of European Americans," Anna C. Need et al., *Genome Biology*, 2009. Available at: http://genomebiology.com/2009/10/1/R7. Using less strict criteria, direct-to-consumer companies such as 23 and Me offer Jewish ancestry testing.

40   The 2001 paper by Harry Ostrer is cited above. For his latest work in this vein, see: "Abraham's Children in the Genome Era: Major Jewish Diaspora Populations Com-

prise Distinct Genetic Clusters with Shared Middle Eastern Ancestry," Gil Atzmon et al., *American Journal of Human Genetics*, Vol. 86, No. 6, June 2010, pp. 850–59.

42   For the dating of the 185delAG mutation, see: "The 185delAG BRCA1 Mutation Originated before the Dispersion of Jews in the Diaspora and Is Not Limited to Ashkenazim," Revital Bruchim Bar-Sade et al., *Human Molecular Genetics*, Vol. 7, No. 5, 1998, pp. 801–5. The senior author and lead investigator, Eitan Friedman, has since refined his estimate of the mutation's age to 2,200 years ago. Source: E. Friedman, personal communication, June 2011.

43   Myriad Genetics, which controls BRCA testing, has statistics on the non-Jewish carriers of 185delAG. Thus: "In the clinical genetic testing setting, a small percentage of individuals who tested positive for the 185delAG mutation were not of Ashkenazi Jewish descent. Of 709 individuals who were identified by Myriad Genetics Laboratories, Inc. (as of the summer of 2002) as carrying the 185delAG mutation in BRCA1, 77 individuals (11%) indicated a non-Ashkenazi ancestry. The remaining 632 individuals indicated Ashkenazi Jewish ancestry." Source: "Identification of Germline 185delAG BRCA1 Mutations in Non-Jewish Americans of Spanish Ancestry from the San Luis Valley, Colorado," Lisa G. Mullineaux et al., *Cancer*, Vol. 98, No. 3, August 2003, p. 600.

44   For the finding of 185delAG in non-Jewish families in Yorkshire, England: "Haplotype and Phenotype Analysis of Six Recurrent BRCA1 Mutations in 61 Families: Results of an International Study," S. L. Neuhausen et al., *American Journal of Human Genetics*, Vol. 58, No. 2, February 1996, p. 275. In June 2011, at a conference in Israel, investigator Eitan Friedman stated that the mutation in England had been analyzed and found to be of independent (non-Jewish) origin. A scientific publication is forthcoming.

45   The ancient Muslim city of Medina is described in *No god but God: The Origins, Evolution, and Future of Islam*, Reza Aslan, Random House, New York, 2005.

46   The Ashkenazim arose a millennium ago from admixture between Jewish pioneers and their European hosts, but the direction of the matings and the fractions of the admixture have been unsettled issues. Research by David B. Goldstein and his colleagues has suggested that Middle Eastern Jewish men mated with local women, e.g., "Founding Mothers of Jewish Communities: Geographically Separated Jewish Groups Were Independently Founded by Very Few Female Ancestors," Mark G. Thomas et al., *American Journal of Human Genetics*, Vol. 70, 2002, pp. 1411–20. Goldstein developed this idea in his book *Jacob's Legacy: A Genetic View of Jewish History*, Yale University Press, New Haven, 2008.

    A contrary scenario emphasizing the founding role of Middle Eastern females was put forward in "The Matrilineal Ancestry of Ashkenazi Jewry: Portrait of a Recent Founder Event," Doron M. Behar et al., *American Journal of Human Genetics*, Vol. 78, March 2006, pp. 487–97. About the population's growth, the authors observed that the "unique, well-documented overall demography [of the Ashkenazim] is consistent with several founding events, repeated bottlenecks, and dramatic expansions, from an estimated number of ~25,000 in 1300 A.D. to >8,500,000 around the turn of the 20th century."

These studies were based on rather limited Y-chromosome and mitochondrial DNA assays. More recently, broad, whole-genome analyses of the Ashkenazim have been undertaken, showing their links to other European, Middle Eastern, and other Jewish populations. See: "Genomic Microsatellites Identify Shared Jewish Ancestry Intermediate Between Middle Eastern and European Populations," Naam M. Kopelman et al., *BMC Genetics*, Vol. 10, No. 8, December 2009 (available at http:// www.biomedcentral.com/1471-2156/10/80); "The Genome-Wide Structure of the Jewish People," Doron M. Behar et al., *Nature*, Vol. 466, July 2010, pp. 238–42; "Abraham's Children in the Genome Era: Major Jewish Diaspora Populations Comprise Distinct Genetic Clusters with Shared Middle Eastern Ancestry," Gil Atzmon et al., *American Journal of Human Genetics*, Vol. 86, No. 6, June 2010, pp. 850–59. (The last paper is by Ostrer and his team.)

The bottom line of the population research: Jews, wherever they may live today, share Middle Eastern ancestry, and for that reason their disparate populations show a high degree of genetic relatedness. An Ashkenazi Jew in Poland is more like a Sephardic Jew in Turkey or Spain than he is like his non-Jewish neighbors in Poland. This is what Harry Ostrer meant when he spoke of genetic threads that were recognizably Jewish.

47 A popular treatment of the Khazar theory of Ashkenazi origins is *The Thirteenth Tribe: The Khazar Empire and its Heritage*, Arthur Koestler, Random House, New York, 1976. See also, more recently, Levy-Coffman, op. cit., and Entine, op. cit., and the rebuff of the theory by Atzmon et al., 2010, op. cit.

49 There are many sources of information on the Jewish genetic disorders. Among those I consulted: *Jewish Genetic Disorders: A Layman's Guide*, Ernest L. Abel, McFarland & Co., Jefferson, NC, 2001; "Prenatal Genetic Screening in the Ashkenazi Jewish Population," Randi E. Zinberg, Ruth Kornreich, Lisa Edelmann, Robert J. Desnick, *Clinics in Perinatology: Metabolic and Genetic Screening*, Vol. 28, No. 2, June, 2001, pp. 367–82; "A Genetic Profile of Contemporary Jewish Populations," Ostrer, op. cit. Up-to-date information can be found at the websites of the Jewish Genetic Disease Consortium, http://jewishgeneticdiseases.org, and of the Israeli National Genetic Database, www.goldenhelix.org/server/israeli, which lists 43 Ashkenazi disorders.

50 This chapter's treatment of race science and Jews benefited from the following: Fishberg, op. cit.; "'The Coefficient of Racial Likeness' and the Future of Craniometry," R. A. Fisher, *Journal of the Royal Anthropological Institute*, Vol. 66, 1936, pp. 57–63; "Geneticists and the Biology of Race Crossing," William B. Provine, *Science*, Vol. 182, 1973, pp. 790–96; *The Mismeasure of Man*, Stephen Jay Gould, W. W. Norton, New York, 1981, 1996; *Defenders of the Race: Jewish Doctors and Race Science in Fin-de-siècle Europe*, John M. Efron, Yale University Press, New Haven, 1995; "The Persistence of Racial Thinking and the Myth of Racial Divergence," S. O. Y. Keita and Rick A. Kittles, *American Anthropologist*, Vol. 99, No. 3, 1997, pp. 534–44; *Statement on 'Race*,' American Anthropological Association, Arlington, VA, 1998 (available at http://www.aaanet.org/stmts/racepp.htm); "Racial Science, Social Science, and the Politics of Jewish Assimilation," Mitchell

B. Hart, *The History of Science Society*, University of Chicago Press, Vol. 90, No. 2, June 1999, pp. 268–97; *Mengele: The Complete Story*, Gerald L. Posner and John Ware, Cooper Square Press, New York, 2000; "Race, Ancestry, and Genes: Implications for Defining Disease Risk," Rick A. Kittles and Kenneth M. Weiss, *Annual Review of Genomics and Human Genetics*, Vol. 4, 2003, pp. 33–67; "If There Are No Races, How Can Jews Be a 'Race'?," Steven Kaplan, *Journal of Modern Jewish Studies*, Vol. 2, No. 1, 2003, pp. 79–96; "The Meaning and Consequences of Morphological Variation," Richard L. Jantz, American Anthropological Association, Arlington, VA, 2004; "In the Name of Public Health—Nazi Racial Hygiene," Susan Bachrach, *New England Journal of Medicine*, Vol. 351, No. 5, July 29, 2004, pp. 417–20; "The Measure of America: How a Rebel Anthropologist Waged War on Racism," Claudia Roth Pierpont, *The New Yorker*, March 8, 2004, pp. 48–63; "Race and Reification in Science," Troy Duster, *Science*, Vol. 307, February 2005, pp. 1050–51; *Studying the Jew: Scholarly Antisemitism in Nazi Germany*, Alan E. Steinweiss, Harvard University Press, Cambridge, 2006; *The Price of Whiteness: Jews, Race, and American Identity*, Eric L. Goldstein, Princeton University Press, Princeton, NJ, 2006; "Human Genetics and Politics as Mutually Beneficial Resources: The Case of the Kaiser Wilhelm Institute for Anthropology, Human Heredity and Eugenics During the Third Reich," Sheila Faith Weiss, *Journal of the History of Biology*, Vol. 39, No. 1, 2006, pp. 41–88.

52 The estimate of 0.5 percent Ashkenazi outbreeding per generation was taken from: "Jewish and Middle Eastern Non-Jewish Populations share a Common Pool of Y-chromosome biallelic haplotypes," M. F. Hammer et al., *Proceedings of the National Academy of Sciences*, Vol. 97, No. 12, June 2000, p. 673. This paper was among the first to establish the Middle Eastern roots of the Ashkenazim.

Maurice Fishberg's finding that Jews were more like their immediate neighbors than they were like Jews in other countries is not contradicted by research, mentioned earlier, showing that Jewish populations share Middle Eastern DNA and that Jews across the Diaspora are more similar to each other than to their hosts. That's because Fishberg compared physical traits or phenotypes, while contemporary scientists compare genotypes, the one not necessarily capturing or corresponding to the other.

54 For Ostrer's important 2010 paper, see the note for page 40.

55 Scientists have conjectured for decades whether carriers of a single Tay-Sachs mutation may have gained an evolutionary advantage through natural selection. The opposing argument is that genetic drift can explain the mutation's spread among Ashkenazi Jews. Thus: "Founder Effect in Tay-Sachs Disease Unlikely," N. C. Myrianthopoulos et al., *American Journal of Human Genetics*, Vol. 24, 1972, pp. 341–42; "Role of Genetic Drift in the High Frequency of Tay-Sachs Disease among Ashkenazic Jews," S. Yokoyama, *Annals of Human Genetics*, Vol. 43, 1979, pp. 133–36; "Heterozygote Advantage in Tay-Sachs Carriers?," B. Spyropoulos et al., *Annals of Human Genetics*, Vol. 33, 1981, pp. 375–80; "Curse and Blessing of the Ghetto," Jared Diamond, *Discover*, Vol. 12, March 1991, pp. 60–66; "Geographic Distribution of Disease Mutations in the Ashkenazi Jewish Population Supports Genetic

Drift over Selection," Neil Risch, Josef Ekstein et al., *American Journal of Human Genetics*, Vol. 72, 2003, pp. 812–22; "The Possibility of a Selection Process in the Ashkenazi Jewish Population," Joel Zlotogora and Gideon Bach, *American Journal of Human Genetics*, Vol. 73, No. 2, 2003, pp. 438–40.

The paper proposing a causal connection among Tay-Sachs carriers, BRCA carriers, and intelligence was "Natural History of Ashkenazi Intelligence," G. Cochran et al., *Journal of Biosocial Science*, Vol. 38, No. 5, September 2006, pp. 659–93 (first published online in 2005). See, for a rebuttal, "Haplotype Structure in Ashkenazi Jewish BRCA1 and BRCA2 Mutation Carriers," Kate Im et al., *Human Genetics*, May 20, 2011 (e-publication ahead of print). The senior author, Bert Gold, started his investigation as a believer in the selective benefits of Jewish BRCA, but his analysis of the mutations changed his mind. Source: B. Gold, personal communication, June 2011.

CHAPTER 4: EL CONVENTO

PHOTO: El Convento, San Luis, Colorado.

61 For a physical inventory (buildings and terrain) of San Luis and its sister villages, see: "The Culebra River Villages of Costilla County, Colorado," prepared by Maria Mondragon-Valdez for the State Historic Preservation Office, Colorado Historical Society, June, 2000. Available at: http://www.coloradohistory-oahp.org/publications/pubs/614.pdf. My cultural overview of Hispanos drew upon *The Hispano Homeland*, Richard L. Nostrand, University of Oklahoma Press, Norman, 1992. The folkways of Culebra are described in *The People of El Valle: A History of the Spanish Settlers in the San Luis Valley*, Olibama Lopez Tushar, El Escritorio Press, Pueblo, CO, 2007, and *La Cultura Constante de San Luis*, edited by Randall Teeuwen, The San Luis Museum and Cultural Center, San Luis, CO, 1985.

62 The area's vermilion hills: *Death Comes for the Archbishop*, Willa Cather, Knopf, New York, 1927, p. 305.

63 About the figure of Doña Sebastiana: Tushar, op. cit., p. 117. References for the *penitentes* are given in chapter 6 notes.

65 For an analysis of the depiction of suffering in Spanish Catholic religious painting, see "Images of Power and Salvation," Rosemary Mulcahy, in *El Greco to Velazquez: Art During the Reign of Philip III*, edited by Sarah Schroth, Distributed Art Publishers, Inc., New York, 2008, pp. 123–45.

66 The genetic proximity of the Sephardim and the Ashkenazim is explored in Atzmon et al., 2010, op. cit.

66 In addition to Potok, Fishberg, and Entine, all op. cit., my treatment of the Jewish and crypto-Jewish experience in medieval Spain benefited from: *The Battle for God: A History of Fundamentalism*, Karen Armstrong, Ballantine, New York, 2001; *To the End of the Earth: A History of the Crypto-Jews of New Mexico*, Stanley M. Hordes, Columbia University Press, New York, 2005; *Juggling Identities: Identity and Authenticity Among the Crypto-Jews*, Seth D. Kunin, Columbia University Press, New York, 2009; "How Muslims Made Europe," Kwame Anthony Appiah, *The New York Review of Books*, November 6, 2008, pp. 59–62, "A Flogging for the Lucky Ones,"

Stefan C. Reif, *The Times Literary Supplement*, Feb. 20, 2009, p. 8; "Modernizing the Marranos," J. H. Elliott, *The New York Review of Books*, March 11, 2010, pp. 22–24.

68 For speculation on the Jewishness of Miguel de Cervantes, see, for one example: "Is There a Hidden Jewish Meaning in *Don Quixote*?," Michael McGaha, *Cervantes: Bulletin of the Cervantes Society of America*, Vol. 21, No. 1, 2004, pp. 173–88.

69 On Jewish genetic markers in Spain and Portugal: "The Genetic Legacy of Religious Diversity and Intolerance: Paternal Lineages of Christians, Jews, and Muslims in the Iberian Peninsula," Susan M. Adams et al., *The American Journal of Human Genetics*, Vol. 83, No. 6, December 2008, pp. 725–36.

70 For the Spanish study of BRCA mutations: "Analysis of BRCA1 and BRCA2 Genes in Spanish Breast/Ovarian Cancer Patients: A High Proportion of Mutations Unique to Spain and Evidence of Founder Effects," Orland Díez et al., *Human Mutation*, Vol. 22, 2003, pp. 301–12.

71 Superstition in the late Middle Ages and state responses to it are discussed in: *Witchcraft and Magic in Europe: The Period of the Witch Trials*, University of Pennsylvania Press, Philadelphia, 2002, and *A History of the Inquisition of Spain*, Henry Charles Lea, Vol. 4, Book 8, Chapter 9, Macmillan, London, 1907 (available at: http://libro.uca.edu/lea4/8lea9.htm).

73 For what the Jehovah's Witnesses think about Job, Satan, and the problem of human sickness, see: "Does the DEVIL Make Us Sick?" at http://www.watchtower.org/e/19990901/article_02.htm, and "When Sickness Is No More!" at http://www.watchtower.org/e/200701/article_03.htm. My edition of the tract Shonnie carried, aka the Golden Book, is: *What Does the Bible Really Teach?*, Watch Tower Bible and Tract Society, Brooklyn, NY, 2006.

76 For Saint Teresa's life, I relied heavily on *Teresa of Avila: The Progress of a Soul*, Cathleen Medwick, Doubleday, New York, 1999. Also of interest are: *The Life of Saint Teresa*, "Taken from the French of 'A Carmelite Nun' by Alice Lady Lovat with a Preface by Mgr. Robert Hugh Benson," Herbert & Daniel, London, 1912; and the nun's own account of her life and spiritual development, *The Life of Teresa of Jesus: The Autobiography of Teresa of Avila*, translated and edited by E. Allison Peers, Image Books (Doubleday), New York, 2004. The "four degrees of prayer," which Teresa likened to watering a garden, are elaborated in *The Life*. And no consideration of Spanish Catholic mysticism is complete without *Tragic Sense of Life*, Miguel de Unamuno, translated by J. E. Crawford Flitch, Dover edition, New York, 1954.

79 For the Jehovah's Witnesses' understanding of the soul, see: "Do You Have an Immortal Soul?" at http://www.watchtower.org/e/20070715/article_01.htm. On the body's immortality: "You Can Live Forever" at http://www.watchtower.org/e/20061001/article_02.htm.

CHAPTER 5: THE LOST TRIBE

PHOTO: Iona and Marianne, San Pablo, Colorado.

90 Along with Nostrand, Tushar, Cather, and Hordes, all op. cit., my sources on the Native American habitation and Spanish settlement of New Mexico are: *Historia de*

la Nueva México, Gaspar Pérez de Villagrá, translated by Gilberto Espinosa, The Quivira Society, Los Angeles, 1933 (excerpts available at http://nationalhumanitiescenter. org/pds/amerbegin/exploration/text1/villagra.pdf); *My Penitente Land: Reflections on Spanish New Mexico*, Fray Angélico Chavez, Museum of New Mexico Press, Santa Fe, 1974; *Santa Fe: History of an Ancient City*, edited by David Grant Noble, School of American Research Press, Santa Fe, 1989; *When Jesus Came, the Corn Mothers Went Away: Marriage, Sexuality, and Power in New Mexico, 1500–1846*, Ramon Gutiérrez, Stanford University Press, Stanford, CA, 1991; "Historical Introduction to North American Missions," John Corrigan, in *French and Spanish Missions in North America*, Electronic Cultural Atlas Initiative, University of California, 2004 (http://www.ecai .org/na-missions/docs/historicalintroduction.htm); *Blood and Thunder: An Epic of the American West*, Hampton Sides, Doubleday, New York, 2006.

90   Regarding the genetic admixture understanding of New Mexico college students: Klimentidis et al., op. cit.

93   Josiah Gregg's commentary on the Spanish conquest is from his book *Commerce of the Prairies, Or, The Journal of a Santa Fe Trader*, J. & H. G. Langley, New York, 1845, p. 267. Like several other out-of-print books in the public domain, it was accessed online at http://books.google.com.

96   About the blend of Native American and European blood in San Luis Valley Hispanos, see "Admixture in the Hispanics of the San Luis Valley, Colorado, and Its Implications for Complex Trait Gene Mapping," C. Bonilla et al., *American Journal of Human Genetics*, Vol. 68, 2004, pp. 139–63. For evidence of one-sided mating in the DNA of SLV Hispanos: "Mitochondrial Versus Nuclear Admixture Estimates Demonstrate a Past History of Directional Mating," D. Andrew Merriwether et al., *American Journal of Physical Anthropology*, Vol. 102, 1997, pp. 153–59.

For the admixture fractions of other Hispanic groups, see: "Genome-Wide Patterns of Population Structure and Admixture among Hispanic/Latino Populations," Katarzyna Bryc et al., *Proceedings of the National Academy of Sciences*, Vol. 107, No. 2, May 2010, pp. 8954–61. Bryc et al. also deal with sex-biased mating, citing the predominantly European Y chromosomes of Hispanic men. Harry Ostrer was senior author on this paper.

98   For the design of the typical New Mexico *plaza*, see: Mondragon-Valdez, op. cit., p. 10.

99   On the stealing of people versus the stealing of animals: Sides, op. cit, p. 128.

99   Sources on the *casta* system of colonial Mexico and New Mexico, in addition to Gutiérrez, op. cit., are: "Españoles, Castas, y Labradores: Santa Fe Society in the Eighteenth Century," Adrian Bustamante, in Noble, 1989, op. cit., pp. 65–77; *Casta Painting: Images of Race in Eighteenth-Century Mexico*, Ilona Katzew, Yale University Press, New Haven, 2004; "Casta Paintings: The Construction and Depiction of Race in Colonial Mexico," Christa Johanna Olson, posted online at: http://hemi.ps.tsoa.nyu.edu/archive/studentwork/colony/olson/Casta1.htm.

For a scientific account of assortative mating, see "Ancestry-Related Assortative Mating in Latino Populations," Neil Risch et al., *Genome Biology*, 2009, posted online at http://genomebiology.com/2009/10/11/R132.

100   For the 2004 SLV admixture study, see Bonilla et al., op. cit. The incident of a man comparing his Indian slave to a Jew is taken from Gutiérrez, op. cit., p. 195.

101 For the historical work of Fray Angélico Chavez, see Chavez, op. cit.

102 That 90 percent of marriages were intravillage is from Gutiérrez, op. cit., p. 282ff.

103 Here are leading references for four genetic disorders that have been recognized in Hispanos: "Oculopharyngeal muscular dystrophy [OPMD] in Hispanic New Mexicans," M. W. Becher et al., *Journal of the American Medical Association*, Vol. 286, No. 19, 2001, pp. 2437–40; "Familial Cavernous Angiomas of the Brain in an Hispanic Family," I. Mason et al., *Neurology*, Vol. 38, No. 2, 1988, pp. 324–26; "Genetic Risk for Recombinant 8 Syndrome [San Luis Valley syndrome] and the Transmission Rate of Balanced Inversion 8 in the Hispanic Population of the Southwestern United States," A. C. Smith et al., *American Journal of Human Genetics*, Vol. 41, No. 6, 1987, pp. 1083–1103. The congenital dwarfism syndrome is reported in "5 Birth-Defect Cases Tell a Centuries-Old Tale," Mindy Sink, *The New York Times*, April 22, 2003. It has not yet been classified, according to one of the investigators, Dr. Carol Clericuzio of the University of New Mexico.

103 For an overview of breast-cancer incidence in Hispanics, including the contribution of European ancestry to women's risk, see: "Genetic Ancestry and Risk of Breast Cancer among U.S. Latinas," Laura Fejerman et al., *Cancer Research*, Vol. 68, No. 23, December 2008, pp. 9723–28.

103 On the genetics of Zuni cystic fibrosis: "Determination of Cystic Fibrosis Carrier Frequency for Zuni Native Americans of New Mexico," D. Kessler et al., *Clinical Genetics*, Vol. 49, No. 2, 1996, pp. 95–97. On the genetics of Zuni kidney disease: "Heritability of Measures of Kidney Disease among Zuni Indians: The Zuni Kidney Project," Jean W. MacCluer et al., *American Journal of Kidney Diseases*, Vol. 56, No. 2, 2010, pp. 289–302.

106 Regarding the herbal remedies of the traditional *médica*, see Tushar, op. cit., p. 84.

109 Dorothy Martinez Medina died in July 2010, following surgery for advanced stomach cancer. The operation, at University of Colorado Hospital in Denver, was performed without a blood transfusion, per the proscription of the Jehovah's Witnesses, and the procedure was said to have been successful. But two weeks later, Dorothy succumbed to pneumonia and heart failure. She had asked not to be kept on life support, another stipulation of the Witnesses.

During the months before her cancer was discovered, Dorothy treated her stomach pain—thought to be ulcers—with traditional remedies rather than the antibiotics that were (mistakenly) prescribed. Was her stomach cancer related to her BRCA1 mutation? The medical literature suggests that BRCA carriers are at greater risk for cancer in general.

Dorothy's sister Bernarda, a breast-cancer survivor, died in July 2011 at the age of ninety-five.

CHAPTER 6: FROM THE *MORADA* TO THE KINGDOM HALL

PHOTOS: *Morada*, San Francisco, Colorado; San Luis Kingdom Hall of the Jehovah's Witnesses, San Pedro, Colorado.

113 On white Americans' repugnance for Hispanos as a race, see Sides, op. cit., pp. 125–26. The quotation about a mongrel race is from *Rocky Mountain Life, or, Startling*

Scenes and Perilous Adventures in the Far West during an Expedition of Three Years, Rufus B. Sage, Wentworth, Boston, 1857, chapter 2. The depiction of Mexican ladies smoking is from *Wah-To-Yah, and the Taos Trail; or Prairie Travel and Scalp Dances, with a look at Los Rancheros from Muleback and the Rocky Mountain Campfire*, Lewis H. Garrard, H. W. Derby & Co., Cincinnati, 1850, p. 196. The racial observations of W. H. H. Davis are taken from his *El Gringo; or, New Mexico and Her People*, Harper & Brothers, New York, 1857, pp. 214ff, 316. Regarding the alleged racial shortcomings of Mexican soldiers: *Types of Mankind, or Ethnological Researches, based upon the Ancient Monuments, Paintings, Sculptures, and Crania of Races, and upon their Natural, Geographical, Philological, and Biblical History*, J. C. Nott, MD, and Geo. R. Gliddon, Lippincott, Grambo & Co., Philadelphia, 1854, pp. 280, 454ff.

115 The quotation by Browne is from his *Adventures in the Apache County: A Tour Through Arizona and Sonora*, Harper & Brothers, New York, 1869, p. 172.

As an example of the higher repute that Americans held for the Pueblo Indians, one Thomas James wrote: "I have spoken before, in favorable terms of the Mexican [*sic*] Indians. They are a nobler race of people than their masters the descendants of the conquerors; more courageous and more generous; more faithful to their word and more ingenious and intellectual than the Spaniards. The men are generally six feet in stature, well formed and of an open, frank, and manly deportment. Their women are very fascinating, and far superior in virtue, as in beauty, to the greater number of the Spanish females." Source: *Three Years Among the Indians and Mexicans*, Gen. Thomas James, Printed at the office of the "War Eagle," Waterloo, IL, 1846. Accessed at: http://www.xmission.com/~drudy/mtman/html/james/jamesint.html.

116 W. H. H. Davis on New Mexicans' "blind adoration": op. cit., p. 225. On his hope for their improvement: p. 231.

116 Willa Cather's comment about the stern religious temperament of the more remote New Mexico villages is from a letter she wrote to *The Commonweal*, explaining how she came to write her novel. The letter was included in later editions of *Death Comes for the Archbishop*.

116 In addition to Mondragon-Valdez and Tushar, both op. cit., two good sources for the tangled history of the Sangre de Cristo Grant are: "Memory and Pluralism on a Property Law Frontier," Gregory Alan Hicks, bepress Legal Series, Working Paper 860, November 14, 2005 (accessible at: http://law.bepress.com/expresso/eps/860), and "Charles Beaubien and Common Use-Rights on the Sangre de Cristo Grant," Malcolm Ebright, Center for Land Grant Studies, Guadalupita, NM, undated MS.

117 For explications of the *penitentes* and their *moradas*, see, in addition to Chavez., op. cit., and Teeuwen, ed., op cit., *Brothers of Light, Brothers of Blood: The Penitentes of the Southwest*, Marta Weigle, University of New Mexico Press, Albuquerque, 1976; and *Land of the Penitentes, Land of Tradition*, Ruben E. Archuleta, Schuster's Printing, Pueblo, CO, 2003.

Details of the restoration of the San Francisco *morada* were provided by Belinda Zink, project architect for the State Historical Fund, Colorado Historical Society.

119 The report of the parading of flagellants in medieval Valladolid comes from Mulcahy, op. cit., p. 135.

121 About the advent of Protestantism in New Mexico and Culebra: *The Protestant Clergy in the Great Plains and Mountain West, 1865–1515*, Ferenc Morton Szasz, University of New Mexico Press, Albuquerque, 1988; *Sea la Luz: The Making of Mexican Protestantism in the American Southwest 1829–1900*, Juan Francisco Martinez, University of North Texas Press, Denton, TX, 2006; "The History of Costilla County as Revealed by Its Cemeteries," Hazel Petty, San Luis Valley Genealogical Society, obtained from http://costillacounty.homestead.com.

122 On the development of premillennial Protestantism from the Great Disappointment to the Jehovah's Witnesses, see: *Mankind's Search for God*, Watch Tower Bible and Tract Society of New York, Inc., Brooklyn, NY, 1990; "The Adventist and Jehovah's Witness Branch of Protestantism," Jerry Bergman, *America's Alternative Religions*, edited by Timothy Miller, State University of New York Press, Albany, 1995, pp. 33–46; "Millions Now Living Will Never Die," Iain S. Maclean, *Religions of the United States in Practice*, edited by Colleen McDannell, Vol. 2, Princeton University Press, Princeton, NJ, 2001, pp. 375–88; *When Prophecy Fails*, Leon Festinger et al., Martino Fine Books, Eastford, CT, 1956, 2009.

   Jehovah's Witness publications, including *The Watchtower* magazine and *The New World Translation of the Holy Scriptures*, can be obtained through any Kingdom Hall or by accessing "Watchtower: Official Web Site of Jehovah's Witnesses," http://www.watchtower.org.

CHAPTER 7: THE DNA AGE

PHOTO: *Santo*, San Luis, Colorado.

138 My treatment of Tay-Sachs and the effort to combat the disease in the Jewish community drew upon: "Screening and Prevention in Tay-Sachs Disease: Origins, Update, and Impact," Michael M. Kaback, *Advances in Genetics*, Vol. 44, 2001, pp. 253–65; "Tay-Sachs Screening in the Jewish Ashkenazi Population: DNA Testing Is the Preferred Procedure," Gideon Bach et al., *American Journal of Medical Genetics*, Vol. 99, 2001, pp. 70–75; "A Genetic Profile of Contemporary Jewish Populations," Harry Ostrer, *Nature Reviews/Genetics*, Vol. 2, November 2001, pp. 895–96; *Jewish Genetic Disorders: A Layman's Guide*, Ernest L. Abel, McFarland & Co., Jefferson, NC, 2001, pp. 126–37; "Using Genetic Tests, Ashkenazi Jews Vanquish a Disease," Gina Kolata, *The New York Times*, February 18, 2003; "Ten Years of Progress," Miriam Colton, *Jewish Daily Forward*, August 20, 2004; "Tay-Sachs Disease," J. A. F. Filho and Barbara Shapiro, *Archives of Neurology*, Vol. 61, 2004, pp. 1466–68. See also the entry on Tay-Sachs in *Online Mendelian Inheritance in Man*, the database maintained by the National Institutes of Health and Johns Hopkins University (http://www.ncbi.nlm.nih.gov/omim/272800). In addition, I interviewed Drs. Michael Kaback, Robert Desnick, and Harry Ostrer about their parts in the TSD screening programs.

140 In addition to my interview with him, my account of Rabbi Ekstein and Dor Yeshorim benefited from: "The Dor Yeshorim Story: Community-Based Carrier Screening for Tay-Sachs Disease," Josef Ekstein and Howard Katzenstein, *Advances in Genetics*, Vol.

44, 2001, pp. 298–310; "The Rabbi's Dilemma," Alison George, *New Scientist*, February 14, 2004, pp. 45–47; "Responsibility of Genetic Testing" (letter), Josef Ekstein, *Genetics in Medicine*, Vol. 6, No. 5, 2004; "Most Studied Yet Least Understood: Perceptions Related to Genetic Risk and Reproductive Genetic Screening in Orthodox Jews," Ilana S. Mittman, PhD dissertation (Philosophy), Johns Hopkins University, Baltimore, MD, 2005; "Trailblazer in Genetics for the Jewish World and Beyond," Yehoshua Leiman, *Personal Glimpses*, supplement to *Hamodia*, Pesach 5766 (April 2006), pp. 24–27.

142   That up to half of Ashkenazi Jews carry at least one harmful mutation is from Dr. Carole Oddoux, a colleague of Harry Ostrer's at NYU School of Medicine. Oddoux offered the estimate at a genetics conference in Israel in 2011.

As for the race to discover the mutation for familial dysautonomia (FD): In 2001, two reports were published in the same issue of the *American Journal of Human Genetics* (vol. 68). Making essentially identical claims, the two papers were: "Familial Dysautonomia Is Caused by Mutations of the IKAP Gene," by Sylvia Anderson et al., and "Tissue-Specific Expression of a Splicing Mutation in the IKBKAP Gene Causes Familial Dysautonomia," by Susan A. Slaugenhaupt et al. Ekstein and Fordham University researchers authored the former paper, while Ostrer, though not an author, was associated with the Massachusetts General Hospital team that produced the latter paper. A patent battle between the two groups, with control of the FD genetic test at stake, was decided in a 2011 court decision in favor of the Massachusetts General team. (The litigation did not involve Ostrer.)

145   The prominence of Jewish subjects in the genetic literature was examined in "Prevalence of Jews as Subjects in Genetic Research: Figures, Explanation, and Potential Implications," Daphna Carmeli, *American Journal of Medical Genetics*, Vol. 130A, No. 1, 2004, pp. 76–83.

146   A good book on the sequencing of the human genome, including explanations of the technology, is: *The Genome War: How Craig Venter Tried to Capture the Code of Life and Save the World*, James Shreeve, Knopf, New York, 2004. See also my own magazine writing on genetics and on the outsized medical hopes for genomics, e.g., "Reading the Language of Our Ancestors," *Discover*, February 2002, pp. 70–77; "Bad Genes, Good Drugs," *Discover*, April 2002, pp. 52–60; and other articles, accessible at http://jeffwheelwright.com.

147   The King–Skolnick competition to isolate BRCA1 is detailed in: *Curing Cancer: Solving One of the Greatest Medical Mysteries of Our Time*, Michael Waldholz, Simon & Schuster, New York, 1997, and *Breakthrough: The Race to Find the Breast Cancer Gene*, Kevin Davies and Michael White, John Wiley & Sons, New York, 1997. The decisive paper from the Skolnick lab was "A Strong Candidate for the Breast and Ovarian Cancer Susceptibility Gene BRCA1," Y. Miki et al., *Science*, Vol. 266, 1994, pp. 66–71.

Skolnick's remark about his joking around with Mormons is from my unpublished interview with him in 2005. King's discovery of her Jewish grandfather is reported by Entine, op. cit., p. 282–89.

148   As to the prevalence and penetrance of 185delAG in Ashkenazi Jews, a number of

reports, studies, and commentaries about the mutation followed quickly on the heels of the identification of BRCA1. See, primarily: "BRCA1 Mutations in Ashkenazi Jewish Women" (letter), Patricia Tonin et al., *American Journal of Human Genetics*, Vol. 57, 1995, p. 189; "A Common *BRCA1* mutation in the Ashkenazim," David Goldgar and Philip Reilly, *Nature Genetics*, Vol. 11, 1995, pp. 113–14; "The Carrier Frequency of the *BRCA1* 185delAG Mutation Is Approximately 1 Percent in Ashkenazi Jewish Individuals" (letter), Jeffery P. Struewing et al., *Nature Genetics*, Vol. 11, 1995, pp. 198–200; "Novel Inherited Mutations and Variable Expressivity of BRCA1 Alleles, Including the Founder Mutation 185delAG in Ashkenazi Jewish Families," Lori Friedman et al., *American Journal of Human Genetics*, Vol. 57, 1995, pp. 1284–97; "BRCA1—Lots of Mutations, Lots of Dilemmas," Francis Collins, *New England Journal of Medicine*, Vol. 334, No. 3, 1996, pp.186–88; "The Risk of Cancer Associated with Specific Mutations of BRCA1 and BRCA2 among Ashkenazi Jews," Jeffery P. Struewing et al., *New England Journal of Medicine*, Vol. 336, No. 20, 1997, pp. 1401–8 (the NIH study).

For a synopsis of Jewish BRCA risk, see "Hereditary Breast Cancer in Jews," Wendy S. Rubinstein, *Familial Cancer*, Vol. 3, 2004, pp. 249–57; and the findings of the New York Breast Cancer Study, cited in the note for page 153.

Because Ashkenazi Jews predominate in Europe and North America, their three founder mutations receive the greatest attention. But from the Israeli perspective, taking into account the Sephardic and Mizrahi populations, there may be as many as seven founder mutations that Jews should recognize. See: "Two BRCA1/2 Founder Mutations in Jews of Sephardic Origin," Michal Sagi et al., *Familial Cancer*, Vol. 10, No. 1, 2011, pp. 59–63.

150 On whether Ashkenazi Jewish women have more breast cancer than other women, a small study conducted in Britain found that they did: "A Population-Based Audit of Ethnicity and Breast Cancer Risk in One General Practice Catchment Area in North London, UK: Implications for Practice," M. Ferris et al., *Hereditary Cancer in Clinical Practice*, Vol. 5, No. 3, 2007, pp. 157–60.

150 The first major study to establish the young age of 185delAG carriers with cancer was: "Germ-Line BRCA1 Mutations in Jewish and Non-Jewish Women with Early-Onset Breast Cancer," M. G. Fitzgerald et al., *New England Journal of Medicine*, Vol. 334, No. 3, 1996, pp. 143–49.

151 For the greater uptake of BRCA screening by Jewish women, see: "Awareness and Attitudes Concerning BRCA Gene Testing," Avigyail Mogilner et al., *Annals of Surgical Oncology*, Vol. 5, No. 7, 1998, pp. 607–12; "Racial Differences in the Use of *BRCA1/2* Testing among Women with a Family History of Breast or Ovarian Cancer," Katrina Armstrong et al., *Journal of the American Medical Association*, Vol. 293, No. 14, 2005, pp. 1729–36.

152 The Ethical, Legal, and Social Implications (ELSI) Research Program of the National Human Genome Research Institute is still operational. For information on past and current projects, go to: http://www.genome.gov/10001618.

Late-1990s concerns by Jews about participating in genetic research are explored in: "Judaism, Genetic Screening and Genetic Therapy," Fred Rosner, *Mt. Sinai Jour-*

nal of Medicine, Vol. 65, Nos. 5 and 6, 1998, pp. 406–13 (available at: http://www
.jewishvirtuallibrary.org/jsource/Judaism/genetic.html); "Toward a Framework
of Mutualism: The Jewish Community in Genetics Research," Karen Rothenberg
and Amy Rutkin, *Community Genetics*, Vol. 1, 1998, pp. 148–53; "Jewish Concern
Grows as Scientists Deepen Studies of Ashkenazi Genes," Sheryl Gay Stolberg, *The
New York Times*, April 22, 1998.

There was evidence that Jewish women avoided BRCA testing because of mis-
placed fears about discrimination in health-insurance coverage. See, for instance:
"Health Insurance and Discrimination Concerns and BRCA1/2 Testing in a Clinic
Population," Emily A. Peterson et al., *Cancer Epidemiology, Biomarkers & Prevention*, Vol. 11, 2002, pp. 79–87.

The ELSI paper complaining that Ashkenazim were targeted by BRCA researchers was: "Ashkenazi Jews and Breast Cancer: The Consequences of Linking Ethnic
Identity to Genetic Disease," *American Journal of Public Health*, Vol. 96, No. 11,
2006, pp. 1979–88.

153 For the findings of the New York Breast Cancer Study: "Breast and Ovarian Cancer
Risks Due to Inherited Mutations in *BRCA1* and *BRCA2*," Mary-Claire King et al.,
Science, Vol. 302, October 24, 2003, pp. 643–45.

154 On the national surge in BRCA testing, see, for one of many examples: "Jewish
Women Change Their Destinies by Testing for Genetic Mutation," Julie G. Fax,
Jewish Journal of Greater Los Angeles, March 27, 2008.

156 For a thorough discussion of prophylactic mastectomy and oophorectomy, see:
*Positive Results: Making the Best Decisions When You're at High Risk for Breast and
Ovarian Cancer*, Joi L. Morris and Ora K. Gordon, MD, Prometheus Books, New
York, 2010. For recent evidence that ovarian removal saves lives: "Association of
Risk-Reducing Surgery in BRCA1 or BRCA2 Mutation Carriers with Cancer Risk
and Mortality," Susan M. Domcheck et al., *Journal of the American Medical Associa-
tion*, Vol. 304, No. 9, 2010, pp. 967–75.

About the relatively high rate of prophylactic mastectomy in the United
States: "Predictors of Contralateral Prophylactic Mastectomy in Women with
a BRCA1 or BRCA2 Mutation: The Hereditary Breast Cancer Clinical Study
Group," Kelly A. Metcalfe et al., *Journal of Clinical Oncology*, Vol. 26, No. 7,
2008, pp. 1093–97.

159 Rubinstein's estimate of the number of lives to be saved from national BRCA
screening for Jews: "Cost Effectiveness of Population-Based BRCA1/2 Testing and
Ovarian Cancer Prevention for Ashkenazi Jews: A Call for Dialogue," Wendy S.
Rubinstein et al., *Genetics in Medicine*, Vol. 11, No. 9, 2009, pp. 629–39.

160 In July 2011 a federal appeals court upheld Myriad's patents on the BRCA genes.
The judicial opinion summarizes the important issues on both sides. See: http://
www.cafc.uscourts.gov/images/stories/opinions-orders/10-1406.pdf.

161 BRCA-associated tumors have a poorer prognosis because they tend to be "triple
negative"—that is, usually the cancer cells lack estrogen or progesterone recep-
tors while evincing normal amounts of HER2 (human epidermal growth factor

receptor-2). As a result, these cancers cannot be attacked by hormone and anti-HER2 therapies. For more information, see: http://www.tnbcfoundation.org/index.html.

As for the frustrating returns from genomics technology, research on type 2 diabetes offers a telling example. Type 2 diabetes is a widespread disorder of high interest to pharmaceutical companies. Years of searching for predisposing alleles (genetic variants) has produced no fewer than forty candidates across the genome, their functions largely unknown. Regardless, these alleles have very minor impact on the development of the condition, whether individually or in combination. If diabetes genes are not going to be able to "explain" diabetes, lifestyle factors such as diet must assume more importance. See: "Genomics, Type 2 Diabetes, and Obesity," Mark I. McCarthy, *New England Journal of Medicine*, Vol. 363, December 2010, pp. 2339–50.

Scientists used to think that the variants revealed by genomics, even if they turned out to be weak, like the diabetes alleles, at least would pertain to large populations of patients, but even that expectation doesn't seem to be true. DNA variation is more variable than anyone knew. Uncommon alleles sequestered in the world's ethnic groups appear to hold the keys to genetic risk. See: "Uncovering the Roles of Rare Variants in Common Disease through Whole-Genome Sequencing," Elizabeth T. Cirulli and David B. Goldstein, *Nature Reviews/Genetics*, Volume 11, June 2010, pp. 415–26; "Genetic Heterogeneity in Human Disease," Jon McClellan and Mary-Claire King, *Cell*, Vol. 141, April 16, 2010, pp. 210–17; "Genomic Medicine—An Updated Primer," W. Gregory Feero et al., *New England Journal of Medicine*, Vol. 362, 2010, pp. 2001–11. The latter two papers contrast the potency of BRCA with the alleles implicated in other diseases. A less technical article about the tardy promise of genomics is: "Awaiting the Genome Payoff," Andrew Pollack, *The New York Times*, June 14, 2010.

162 A frequently cited critique of DTC gene testing is: "Direct-to-Consumer Genetic Tests: Misleading Test Results Are Further Complicated by Deceptive Marketing and Other Questionable Practices," U.S. Government Accountability Office (GAO), July 22, 2010, available at: http://www.gao.gov/products/GAO-10-847T. A more sanguine review of "recreational genomics" is: "My Genome, My Self," Steven Pinker, *The New York Times Magazine*, January 11, 2009, available at: http://www.nytimes.com/2009/01/11/magazine/11Genome-t.html. For more on the dozen "personal genome pioneers" who gathered in Cambridge, Massachusetts, in 2010, go to: http://www.getconference.org. See also: "Vanity Genomes and the Future of Medical Sequencing," National Human Genome Research Institute, September 2010 (http://www.genome.gov/27527308).

163 As to MRI (magnetic resonance imaging) screening for high-risk women: Because of their generally younger ages and denser breasts, and because their tumors, if not caught early, can be harder to treat, BRCA carriers are advised to have MRI scans along with or instead of mammograms. For these patients, the sensitivity of MRI is about twice that of mammography, according to several studies, although the cost

of MRI can be a deterrent. See: "M.R.I.'s Help Fight High Risk of Cancer," Denise Grady, *The New York Times*, November 15, 2010, and Morris and Gordon, op. cit., pp. 186–94.

CHAPTER 8: LAST DAYS OF THE INDIAN PRINCESS

PHOTO: Shonnie, Alamosa, Colorado (credit: George Casias).

169 For information and respectful advice about alternative medicine, I recommend the website of the National Center for Complementary and Alternative Medicine, a unit of the National Institutes of Health, at http://nccam.nih.gov.

CHAPTER 9: WHEN HARRY MET STANLEY

PHOTO: DNA sampling session, San Pablo, Colorado.

184 The first published report of 185delAG in Hispanos was: "Identification of Germline 185delAG BRCA1 Mutations in Non-Jewish Americans of Spanish Ancestry from the San Luis Valley, Colorado," Lisa G. Mullineaux et al., *Cancer*, Vol. 98, No. 3, August 2003. I first learned about the mutation from a talk given by Mary-Claire King in November of that year. King had come across 185delAG in a large Hispano family named Trujillo during the late 1990s but had not reported it formally.

Sharon Graw, a contributor to the 2003 *Cancer* paper, performed a molecular analysis of the Hispano mutation and confirmed its Jewish origin. See: "BRCA1:185delAG Found in the San Luis Valley Probably Originated in a Jewish Founder," I. Makriyianni et al., *Journal of Medical Genetics*, Vol. 42, 2005, e27 (available at http://www.ncbi.nlm.nih.gov/pmc/articles/PMC1736052/pdf/v042p00e27.pdf).

188 Hordes published a number of papers on crypto-Judaism in New Mexico, but his thesis is laid out most fully in his book *To the End of the Earth: A History of the Crypto-Jews of New Mexico*, Stanley M. Hordes, Columbia University Press, New York, 2005, from which I have quoted here.

190 See Chavez, op. cit., for his Hebrew–New Mexico analogy.

191 Two early works by Judith Neulander that disputed Hordes are: "Crypto-Jews of the Southwest: An Imagined Community," *Jewish Folklore and Ethnology Review*, Vol. 16, No. 1, 1994, pp. 64–68, and "The New Mexican Crypto-Jewish Canon: Choosing to be 'Chosen' in Millennial Tradition," *Jewish Folklore and Ethnology Review*, Vol. 18, Nos. 1–2, 1996, pp. 19–58.

192 The magazine article supporting Neulander was: "Mistaken Identity?: The Case of New Mexico's 'Hidden Jews'," Barbara Ferry and Debbie Nathan, *The Atlantic Monthly*, Vol. 286, December 2000, pp. 85–96. Seth Kunin's book is cited in the note for page 66.

193 About pemphigus vulgaris in New Mexico: "The Historical and Geomedical Immunogenetics of Pemphigus Among the Descendants of Sephardic Jews in New Mexico," Kristine Bordenave et al., *Archives of Dermatology*, Vol. 137, 2001, pp. 825–26. See also Hordes, op. cit, Appendix, pp. 289–95. About Bloom syndrome in New

Mexico: "The Ashkenazic Jewish Bloom Syndrome Mutation blm^{Ash} Is Present in Non-Jewish Americans of Spanish Ancestry," Nathan A. Ellis et al., *American Journal of Human Genetics*, Vol. 63, 1998, pp. 1685–93, and Hordes, op. cit., pp. 271–72.

194 For two examples of the extensive reporting on Father Bill Sanchez, see: "Hispanics Uncovering Roots as Inquisition's 'Hidden' Jews," Simon Romero, *The New York Times*, October 29, 2005, and "The Secret of San Luis Valley," Jeff Wheelwright, *Smithsonian*, October 2008, pp. 48–56.

194 The genetic study that Neulander coauthored was: "Toward Resolution of the Debate Regarding Purported Crypto-Jews in a Spanish-American Population: Evidence from the Y chromosome," Wesley K. Sutton et al., *Annals of Human Biology*, Vol. 33, January–February 2006, pp. 100–111. In arguing that Hispano males were not different from Iberian males when markers on their Y chromosomes were compared with those of Jews, the authors did not examine the possibility that Sephardic Jewish markers had previously been incorporated in Iberians and by extension New Mexicans, per the admixture study by Adams et al., op. cit.

Neulander's complaints about pseudoscience are from "Folk Taxonomy, Prejudice and the Human Genome: Using Disease as a Jewish Ethnic Marker," Judith S. Neulander, *Patterns of Prejudice*, Vol. 40, Nos. 4–5, 2006, pp. 381–98. The pemphigus research that she cited was: "Common Ancestral Origin of Pemphigus Vulgaris in Jews and Spaniards: A Study Using Microsatellite Markers," R. Loewenthal et al., *Tissue Antigens*, Vol. 63, No. 4, pp. 326–34.

196 Regarding the adjustments made to the Cohan Modal Haplotype, compare and contrast the following papers. The first introduced the CMH concept and the second, by the same team, revised it with new data: "Y Chromosomes of Jewish Priests," Karl Skorecki et al., *Nature*, Vol. 385, 1997, p. 32, and "Extended Y Chromosome Haplotypes Resolve Multiple and Unique Lineages of the Jewish Priesthood," Michael F. Hammer et al., *Human Genetics*, Vol. 126, 2009, pp. 707–17.

197 Given the well-documented Ashkenazi presence in New Mexico, it might well be asked if this group was the source of 185delAG in Hispanos, rather than the shadowy Spanish *conversos*. The major argument against the Ashkenazi scenario is that if Eastern European Jews had spread BRCA genes into the local populace, investigators would have detected not only 185delAG but also the other two mutations from the Jewish set: 5382insC in BRCA1 and 6174delT in BRCA2. Together, they are more common in Ashkenazi Jews than 185delAG. Though a rigorous epidemiologic survey is lacking, the other two mutations haven't turned up in clinical samples in New Mexico and Colorado Hispanos. Nor have they been reported in Spain's population, although 185delAG has—which is just what you would expect if 185delAG's transmission was exclusively from Spain. Sources: Mullineaux, op. cit., and Paul Duncan, Hematology-Oncology Associates, Albuquerque, NM, personal communication.

See also, on this question: "Prevalence of BRCA Mutations and Founder Effect in High-Risk Hispanic Families," Jeffrey N. Weitzel et al., *Cancer Epidemiology, Biomarkers & Prevention*, Vol. 14, No. 7, 2005, pp. 1666–71. This study of Hispanics in the Los Angeles area uncovered four instances of 185delAG; none of 6714delT;

and one of 5382insC, in a person who had both Mexican and Eastern European ancestry.

198 An account of the kosher beef project in the San Luis Valley appears in "Hard Choices: The Birth and Death of Ranchers' Choice Cooperative," David Carter, *New Generation Cooperatives: Case Studies*, Illinois Institute for Rural Affairs, Macomb, IL, 2001, pp. 121–31.

202 The newspaper article about Hispano genetic disorders in San Luis Valley, featuring Beatrice, Shonnie, and other patients, was: "Valley Harbors a Tragic Legacy," Karen Augé, *The Denver Post*, September 28, 2003, p. A1. The case of Beatrice Martinez Wright subsequently was taken up in Entine, op. cit., chapter 8. I learned of Shonnie Medina's case during a reporting trip for *Smithsonian* magazine in 2007. I included Beatrice but not Shonnie in my article (Wheelwright, op cit.).

205 For references on the race concept in science, see notes for chapter 3. In addition, the following articles cover the recent influence of genetics on race: "Genetic Structure of Human Populations," Noah Rosenberg et al., *Science*, Vol. 298, December 2002, pp. 2381–85; "Deconstructing the Relationship between Genetics and Race," Michael Bamshad et al., *Nature Reviews/Genetics*, Vol. 5, August 2004, pp. 598–609; "What We Do and Don't Know about 'Race,' 'Ethnicity,' Genetics and Health at the Dawn of the Genome Era," Francis S. Collins, *Nature Genetics*, Supplement, Vol. 36, No. 11, 2004, pp. 513–15; "Genes, Race, and Medicine," Jeff Wheelwright, a three-part series, *Discover*, March, April, May 2005; "Looking for Race in All the Wrong Places: Analyzing the Lack of Productivity in the Ongoing Debate about Race and Genetics," Morris W. Foster, *Human Genetics*, Vol. 126, No. 3, 2009, pp. 355–62.

206 The estimate that twenty-four million base pairs differentiate one person's DNA from another's was taken from Feero et al., op. cit., p. 2003.

207 Regarding the technology of ancestry-testing, direct-to-consumer companies such as 23andMe and Family Tree DNA have started to incorporate more expensive, whole-genome assays in their offerings (i.e., autosomal SNP coverage), the same technology that Ostrer and other scientists use. This is in addition to the standard Y chromosome and mtDNA assays.

207 That DNA can confirm self-reports of racial and ethnic identity was neatly demonstrated by "Genetic Structure, Self-Identified Race/Ethnicity, and Confounding in Case-Control Association Studies," Hua Tang et al., *American Journal of Human Genetics*, Vol. 76, 2005, pp. 268–75.

208 The research published by the Ostrer team on the ancestry of Hispanic groups was: Bryc et al., op. cit.

210 For the *Pueblo Chieftain*'s coverage of the DNA session, see the article by Matt Hildner posted online on February 22, 2009, at http://bit.ly/9VA99V.

215 For descriptions of Laron syndrome, see: "Diverse Growth Hormone Receptor Gene Mutations in Laron Syndrome," Mary Anne Berg et al., *American Journal of Human Genetics*, Vol. 52, 1993, pp. 998–1005, and "Growth Hormone Receptor Deficiency in Ecuador," Arlan Rosenbloom et al., *Journal of Clinical Endocrinology & Metabolism*, Vol. 84, 1999, pp. 4436–43.

On the history of crypto-Judaism in Peru and Ecuador: *La Herencia Sefardita en la Provincia de Loja*, Ricardo Ordóñez-Chiriboga, Casa de la Cultura Ecuatoriana "Benjamin Carrión," Quito, Ecuador, 2005 (translated by Christopher Velez, personal communication).

216 Harry Ostrer's study of the Jewishness of Hispano (and Ecuadorian) DNA is "The impact of *Converso* Jews on the Genomes of Modern Latin Americans," C. Velez et al., *Human Genetics*, DOI 10.1007/s00439-011-1072-z (online publication), July 26, 2011.

CHAPTER 10: THE OBLIGATE CARRIER

PHOTO: Joseph descending, San Luis Valley, Colorado.

222 The Cather quote is from *Death Comes for the Archbishop*, op. cit.

223 My account of the litigation over access to *la sierra* is drawn largely from an interview with attorney Jeffrey Goldstein. See also: "A Little Cloud on the Title," Calvin Trillin, *The New Yorker*, April 26, 1976, pp. 122–32; Teeuwen, ed., op. cit., pp.

INDEX